# FERRARI
# 250GT SWB

# About the Author

Among many other contributions to automotive
journalism and authorship, Ken Gross has been
Feature Editor for *Special Interest Autos* and
Automotive Editor for *Gentleman's Quarterly*.
He is a previous winner of the prestigious
Cugnot Award for automotive history – and
a member of the Madison Avenue Sports Car
Driving and Chowder Society! Like all of the
contributors to the Osprey Expert History
series, he is one of that elite band of writers
who not only know the subject intimately,
but also know how to bring technical
information to life.

# FERRARI 250GT SWB

## KEN GROSS

First published in Great Britain in 1985 by Osprey Publishing,
Elms Court, Chapel Way, Botley, Oxford OX2 9LP

Revised edition published Summer 1999

ISBN 1 85532 884 4

Editors: Tim Parker and Shaun Barrington
New colour section design: The Black Spot
Jacket photography by David Sparrow
Photography for second edition by David Sparrow

The Publishers of the second edition would like to express their
sincere thanks to the owners of the cars included for the first time
in the colour section: Martin Lange, Clive Beecham and Tony Dyes.
They were happy to provide their priceless cars for photography,
when contacted via the Ferrari Owner's Club of Great Britain,
to whom the Publishers also extend their thanks.

New colour section origination: Valhaven Ltd, Isleworth, UK
Printed through Worldprint Ltd, Hong Kong

99  00  01  02  03    10  9  8  7  6  5  4  3  2  1

**Other titles published by Osprey Automotive include:**

*MG by McComb*, third edition updated by Jonathan Wood
    ISBN 185532 831 3
*Jaguar E-Type* , Denis Jenkinson
    ISBN 185532 881 X
*Ferrari F40,* David Sparrow
    ISBN 185532438 5
*Ferrari 512TR,* David Sparrow
    ISBN 185532 439 3
*Mercedes-Benz SL & SLC,* L.J.K. Setright
    ISBN 185532 880 1

For a catalogue of all books by Osprey Military, Aviation and
Automotive please write to:

**The Marketing Manager, Osprey Publishing Ltd.,**
**P.O. Box 140, Wellingborough, Northants NN8 4ZA**

# Contents

Introduction  *6*

Chapter 1  **Background**  *10*

Chapter 2  **The LWB interim berlinettas**  *19*

Chapter 3  **Design and development**  *25*

Chapter 4  **SWB engines**  *45*

Chapter 5  **Coachwork**  *56*

Chapter 6  **Special** *carrozzerie*  *74*

Chapter 7  **Racing highlights**  *88*

Chapter 8  **What's it like to drive?**  *99*

Chapter 9  **SWB successors**  *111*

Specifications  *116*

Production record  *117*

Competition  *121*

Special-bodied SWBs  *124*

Acknowledgements  *125*

Index  *126*

# Introduction

The inspiration for this book was a conversation a while back with Lowell Paddock, the senior editor of *Automobile Quarterly*. We were discussing candidates for the finest dual-purpose sports car of all time and, despite a number of contenders, it didn't take long for us to reach a consensus. Both our votes were cast for the Ferrari 250GT Short-Wheelbase Berlinetta, better known as the 250GT SWB.

After all, here was the quintessential 'drive-it-to-the-track, race it and drive it home' car. The SWB was winning races back when a GT contender had to be a tractable road car, not yet a thinly-disguised out-and-out

*Still beautiful after all these years. Here's 3233 GT, an exquisitely restored steel-bodied Lusso owned by Gary Schaevitz. Photo Alan Boe*

racer. Stories are legion, of SWB owners showing up at a race meet, taking the laurels and heading for home. All in a day's work for this remarkably consistent coupé.

One of my favourite stories illustrating the SWB's capabilities comes from Ferrari historian, Gerald Roush: 'In the early 1960s, at a race on an abandoned air field in Alabama,' Roush wrote, 'we watched a better-than-

*A right-hand drive berlinetta, 3657 GT, exhibits its aggressive egg crate grille. Smooth fender lines contrast with but don't contradict the squared hoodscoop. Photo Alan Boe*

*Scarlet paint, timeless curves, and four big tailpipes let you know you've been passed by something very, very fast. At rest, here's 4065 GT, the last SWB built. Outside filler cap has given way to a smoothly integrated fender flap. Photo Alan Boe*

7

*Here's Enzo Ferrari driving a late SWB. Universally considered to be one of Ferrari's best efforts, the timeless design and outstanding performance of these cars is now, like the man himself, the stuff of legends. Photo the late Peter Coltrin, courtesy Dean Batchelor*

average driver run away from the competition in a gold 250 Testa Rossa. The TR had arrived at the track on a trailer, while the driver had come in a matching gold 250GT SWB loaded with the driver's personal luggage.

'However, the TR developed an ailment of some type on Saturday, and was withdrawn from Sunday's main event. The driver merely took his luggage out of his SWB, and entered his street machine in the feature race, although this meant starting from the rear of the grid. We can't remember for sure whether he won the race or not, but he was certainly the class of the field and the fastest car there as he moved quickly through the pack to run with the leaders. That, we told ourselves, was what a Ferrari was all about. Equally at home on the street or on the race course.'

Gerald's story appeared in a late October, 1977, issue of his *Ferrari Market Letter*. At that time, SWBs were trading at the $20–25,000 level. Today, with prices for SWBs twenty/thirty times higher, there is a reason. Marc Tauber, noted Ferrari broker, has his answer. 'There's a lot of hype on SWBs today,' he says. 'They were true dual-purpose cars and they've still got very close to modern performance. Why do they bring top dollar? They're 'affordable' competition cars, by Ferrari standards; they look right and the factory didn't make too many. Add it up, and you've got all the factors for a fast-appreciating Ferrari.'

I certainly agree. The Tour de France berlinettas were too much of a race car; the fabulous GTOs – not even *thinly* disguised racers – are virtually unattainable. In between, Ferrari built an extraordinary GT. You could race it or run it on the street in various configurations of alloy and steel bodies, big carbs and valves or more restrained manifolding, spartan cockpits or quite luxurious fittings. And, most importantly, some people raced them successfully *and* ran them on the street as well. Little wonder then, why today nearly every SWB has been painstakingly tracked down (although there are a few out there left to be discovered) and they're enjoying a resurgence of interest unparallelled in Ferrari collecting.

Need convincing? A few minutes behind the wheel of an SWB will have you believing you've somehow travelled back in time. The poplar-lined *Route Nationale* stretches in front, the tach needle is straining for the 8000 rpm mark, the wonderful shriek of the 3-litre V12 rings in our ears and you head the quick little coupé for the horizon. You can't hear your co-driver's comments over the noise but it doesn't matter; the years have disappeared, it's 1961, you know where you're going – and you know nothing on wheels can catch you. . . .

Ken Gross
Princeton, New Jersey, USA

# Background

Below *The 250 Europa GT began at the 1953 Paris Salon. Here's a 1956 version. The sweep of the fenders, low roofline and many body details are already beginning to predetermine the SWB*

Above right *The long-wheelbase California Spyder (1958–60) used the 250 engine in a lightweight, very attractive open car. Shown is 1505 GT. Photo Marc Tauber*

Below right *The short-wheelbase California, a later development built from 1960–63, used the updated SWB coupé improvements. The Ferrari nameplate on this car's hood is incorrect as are the wheels and fender mirrors. Photo Marc Tauber*

The term 'berlinetta' literally means, 'little coupé' in Italian, but it's come to connote a racing coupé – particularly where Ferraris are concerned.

Compared with Ferrari's first street cars, the berlinettas were usually lighter, more spartan, and noticeably quicker. In the interest of function, plexiglass replaced glass rear and side windows, alloy bodies were *de rigueur*, and such niceties as soundproofing and bumpers were frequently omitted. Air scoops for engines and brakes were often cut into appropriate surfaces.

Engines, as befitting a racing car, benefited from a higher state of tune – bigger carburettors, lightweight

pistons, hotter cams and higher compression ratios were usually found and before clutch-governed fans, even this appendage was dropped for an extra 2–3 bhp.

Although most Ferrari coupés have been called berlinettas, the cars that really earned the name, the Ferrari 250GTs, began at the 1954 Paris Salon with the Europa, followed by the series-built Boano low and high-roof coupés. By 1956, a more definitive 250GT had emerged and its success in dominating the ten day French road racing classic, led to a new name, the Tour de France.

Along the way, 250GT spyders and cabriolets were built in limited numbers. The Series I Pinin Farina Spyders are lovely roadsters with many unique features – with bodies designed and built by Pinin Farina for the rich and famous of the era. The Aga Khan, Porfirio Rubirosa, and Count Volpi, the racer, were just a few of the early clientele for these exquisite cars.

At his point in the open car history, a branch occurred as the Series II Pinin Farina 250 Cabriolet became the luxurious, sporty street convertible, while a stripped, racier version, the California Spyder, was developed as a competition car. These Spyders, although Pinin Farina designs, were built by Scaglietti on the 2600 mm long-

*Far left Bill Harrah's 1958 LWB 250 Tour de France, 0925 GT, is a beautiful, low-mileage example. In contrast to this model's racing history, noted car collector Harrah used his TDF as a street car, taking advantage of Nevada's unlimited speeds at the time. Looking closely at this car and the California Spyders, it's easy to spot a number of common design features. Photo Dean Batchelor*

*Another variation on the TDF theme, here's a lightweight Zagato-bodied berlinetta. Bumperless body, scooped and stacked 250 engine all meant business. These cars were successfully campaigned by privateers in Europe. 'Double bubble' roof was a Zagato trademark. Photo Karl Dedolph*

*TDF's were the scourge of the competition during their era. Though handling and chassis development lagged behind engine improvements, the quick coupés took the measure of most of their competition. Photo Karl Dedolph*

*Early TDF's featured enclosed lamps, characteristic Ferrari grille with driving lamps, finned rear fenders. Photo Karl Dedolph*

wheelbase Ferrari chassis, at first – and some of the cars were equipped with modified engines, external racing filler caps and lightweight alloy coachwork. The last of the LWB Californias were the most desirable of the series as they had four-wheel disc brakes, reworked, updated suspensions and the new type 128 DF outside plug engine. Styling details are reminiscent of the LWB TDF coupés.

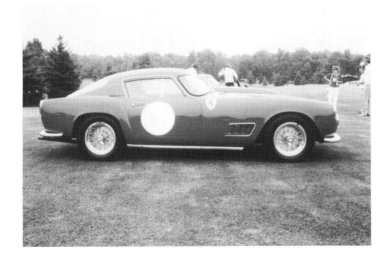

Left and below left 1357 GT *had an interesting hillclimb competition history. It's typical of the 1959 TDFs with open headlamps. The enclosed lights were superseded by a need to conform to a new Italian lighting regulation, but for cars bound to other countries, covered lamps could still be specified. Photos Karl Dedolph*

In March, 1960, the short-wheelbase (2400 mm) version of the California bowed in at the Geneva Salon. This shorter, wider car had many of the SWB coupé's additions – tubular shocks, the new high output, outside plug engine and Testa Rossa type heads. Styling was a carryover from the LWB Californias and did not resemble their SWB coupé counterparts.

Some of the Californias had distinctive racing histories. *1451 GT*, an LWB example, with an alloy body and a centred outside filler cap, was driven to a fifth place at Le Mans in 1959 by Bob Grossman and Fernand Tavano.

Leaving open cars for a moment, let's pick up the parallel history of the berlinettas once again. After the terrible Le Mans accident in 1955, the FIA established a class for Grand Touring cars (a move away from increasingly powerful open sports racers) and Pinin Farina's designs, as they had with the roadsters, became Enzo Ferrari's choice for his new enclosed cars. (Both the man, and the company changed names after 1959 – from Pinin Farina to Pininfarina.)

*The racing background is Sebring, 1958. Curious paint theme doesn't detract from TDF's voluptuous lines. Photos Karl Dedolph*

The Tour de France berlinettas, as the LWB 250GTs came to be called, went through a progressive series of modifications, becoming faster and sleeker each year. Ferrari's coupés had the legs on the competition with

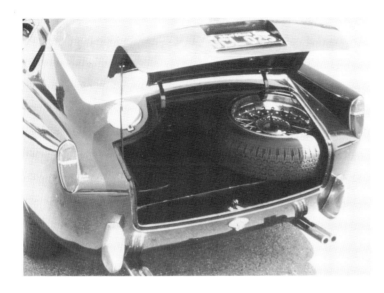

their superior high-output V12 engines. Though chassis development, handling and braking development tended to lag behind, the fiery red cars from Maranello usually prevailed against the opposition.

On the track or in road events, TDFs frequently took top honours from 1956 until the SWBs bowed in late 1959. Built by Scaglietti in Modena, following the Farina design specifications, the TDFs were essentially a continued development of the original berlinetta concept – with light alloy bodies, first open, later enclosed headlights, cold-air boxes, and pronounced cooling vents. While they were frequently used as street cars, the TDF's cramped cockpit, poor ventilation, and lack of soundproofing – all perfectly acceptable for racing – made them cars suitable only for the most enthusiastic drivers.

As more and more customers demanded dual-purpose cars, Ferrari worked on the TDF's successor. He planned a true berlinetta, of course, and he also envisioned a more luxurious version of the same car that could still be competitive in club racing events – and serve its owner well in both modes.

They still race 'em! TDF strikes a calm pose at vintage race meet. While a little underpowered and lacking the handling of later cars, the TDF can still provide a lot of thrills at a GT revival – not to mention all the fabulous sounds of a V12 being driven in anger. Photo Karl Dedolph

Ferrari's plans for his new car would begin with a test of some of its features, at Le Mans. Observers of the 1959 classic may have guessed that the newly reshaped Ferrari berlinettas would be a preview of an even greater car that was soon to come.

# CHAPTER 2

# The LWB interim berlinettas

The SWB had its immediate origins in a very limited production predecessor. In mid-1959, after completing only 13 first series TDF berlinettas, Ferrari built an even shorter series of seven cars on the long-wheelbase Tour de France chassis in order to test, at high speeds, the forthcoming new berlinetta's body design.

The first competition appearance of the Pininfarina

*The LWB 'interim' Berlinetta, a brief, but important step on the way to the eventual SWB. Only seven cars with this configuration were built, in mid-1959, primarily so Ferrari could test the new bodywork prior to the SWB's fall introduction*

prototype was at Le Mans in 1959, accompanied by a Scaglietti-built berlinetta lookalike. These seven cars were unofficially called 'interim' berlinettas by Dick Merritt and Warren Fitzgerald, (the Ferrari factory never officially used this designation) as the cars were a styling link between the old TDF coupés and the new SWB in its initial configuration, scheduled for an autumn 1959, Paris introduction.

Typically for Ferrari, and uniquely unlike most manufacturers, if Enzo Ferrari wanted to tool up for a brief production run of cars to experiment a little with aerodynamics, he'd simply decide to do it. As Jess G. Pourret commented in the *250 GT Competition*, 'The costs of running a smaller, but well organized plant allowed for such a move. What stands between Mr. Ferrari's orders and the drawing boards or the machine tools is a few yards of hallway. When other factories start to think about modifying something, and get the responsible people together after numerous meetings, tons of paper and unmeasurable inertia to finally get the changes underway, Ferrari has long ago tested the end result and, if satisfied, has already incorporated the new ideas into the product. At least in those days of highly skilled labour and union quietness, that's how it worked.'

The new, softer body shape was characterized by small, angular rear quarter windows, a fastback similar to the earlier TDFs and a longer, more sculpted and extended nose and front fenders. The hood incorporated a low, flat-topped airscoop and some of the early cars also had non-factory bug deflectors in front of the driver.

The Pininfarina prototype, *1377 GT*, had front quarter windwings, but these were omitted on the Scaglietti car – in favour of lightweight sliding perspex windows. The Scaglietti prototype, *1461 GT*, also lacked a hoodscoop and bumpers, but it retained the rather unattractive vestigial rear windows which would be omitted on the shorter, later cars. Big fog lamps filled the eggcrate grille's corners – and on some cars a prancing horse was mounted in the centre.

Stepping back from an LWB interim TDF car today and, overlooking the longer wheelbase, it's not hard to see the beginnings of the smooth, purposeful lines that hallmarked the SWBs – all the clues are there for the viewer.

The SWB's blunt front end, however, didn't help the attractive new car achieve its planned higher speed potential. Curiously, this same treatment was carried over to succeeding SWBs. Although Ferrari only began seriously using aerodynamics with the 250GTO – and that car's sleek, extended nose was efficient at high speeds – curiously a similar error appeared with the first series of 250LMs and 275GTBs with Pininfarina (the name changed from Pinin Farina after 1959) taking two tries with short 'noses' before achieving a long-nosed frontal aspect that finally aided his berlinettas at top speed.

Small air scoops for brake cooling appeared at the inner front fender contours. Later, on the SWBs, this

*Symbolically ahead of a TDF at Laguna Seca, Karl Dedolph's impressively restored 1521 GT LWB car looks a great deal like the later SWBs from this angle. Photo Karl Dedolph*

opening would be neatly sleeved. An airvent was also built into the roofline, just above the plexiglass rear window, to help exit cockpit air.

The massive Monza-type racing filler cap was initially located in a notched left hand corner of the decklid. Later SWB versions saw the filler cap recessed into the top curve of the left rear fender.

Inside, a functional black crackle-finish dash was provided, with the instruments grouped handily behind a slightly dished wood-rimmed wheel. Leather or Everflex vinyl bucket seats, with a passenger side headrest, straddled a typically Ferrari tall gear-lever topped with a comfortable alloy ball for the frequent use this component demanded. Most competitor cars had functional rubber mats – a few had carpets.

Under the hood, for five of the seven interim TDF cars, a new engine was provided. Ferrari called it the 128DF/128F and it came with the newly adopted 12-port induction outside plug cylinder heads. However, two interim LWBs were equipped with the older 128D inside plug engine. Karl Dedolph reports this may have happened because of a temporary shortage of the new engines. Olivier Gendebien told him this, in order to

*The curious rear window shape of the LWB 'interim' coupé disappeared when the eventual SWB design was finalized. Lack of cockpit space precluded the extra view – no matter, as it's not the most attractive element of this design. Photo Alan Boe*

have extra engines for some equipe, the motors were taken from cars under construction. On the new power plants, moving the spark plugs outside the cylinder vee lessened the chances of fire, and the outside plugs were easier to change in the frenzy of a race.

Minor differences distinguished the two Le Mans coupés: they used different experimental Weber carburettors – 40 DCZ 6 and 40 DCZ 3 – and 8.5:1 and 9.6:1 compression ratios. The cars shared the same magnesium alloy sumps, four-speed gearboxes and rear end ratios, but the PF prototype berlinetta had an offset gear lever with a reversed shift pattern. All the cars in this short series inherited the earlier TDF's 508D chassis in largely unchanged form. The specifications included large finned drum brakes, a limited-slip differential, Houdaille lever shocks and Borrani RW 3264 16 in. wire

*Viewed from the front, 'interim' Berlinetta looks very much like the later SWB, complete with vestigial bumperettes. Le Man racers used an aluminium plate to regulate temperature/air flow. Scaglietti body to Pininfarina design became the format for SWB's as well. Still ready to compete, Dedolph's 'interim' Berlinetta has racing bug screen in front of driver, squat, purposeful stance. Photo Karl Dedolph*

wheels. Karl Dedolph advised that competition wire wheels were painted various shades of grey in order to distinguish one team's wheels from another at the tyre distributor's stand – hence the difficulty in 'matching' any correct paint today. Tyres were sometimes Englebert but usually Dunlop with a racing compound.

None of the LWB interim cars were ever factory fitted with disc brakes. Ferrari was slow to adopt this new technology. Jess Pourret noted that some owners later retrofitted their berlinettas with Campagnolo discs – to match the Dunlop-braked SWBs – and two cars, *1461 GT* and *1523 GT*, (the last interim berlinetta), were equipped with Dunlop disc brakes for the 1959 Tour de France.

Looking closely at the 1959 Le Mans cars, the grille is fitted with an aluminium plate designed to restrict the air flow for the cooler night portion of the 24-hour classic – and to open the grille intake area during the warmer daylight hours. In the race, the interim cars ran quite well, but the GT winner (and third overall) was a regular TDF berlinetta (1321 GT) with the standard 128D engine.

The interim cars marked the end of the very successful 250GT LWB series. When the short-wheelbase models appeared, competition drivers clamoured for the first few cars and by early 1960, the interim berlinettas were no longer competitive. They became quick street cars for a few lucky owners.

Today, they're an interesting dead end in Ferrari history, very rare and certainly recognized by Ferrari historians as a key step toward the creation of a car that's popularly thought to be the best all-round track and street sports car of all time.

# Design and development

The first short-wheelbase berlinetta was launched, as Ferrari so often liked, at the October, 1959, Paris Salon. Hastily prepared, as Scaglietti was preoccupied with the last of the interim cars, the initial SWB, as it quickly became known, arrived in Paris with drum brakes. Dunlop discs were quickly installed prior to the show's opening.

The prototype, *1539 GT*, was dark red with a natural leather interior. The first car featured an alloy body and was equipped with bumpers. Neither air vents nor side

*Taken from the first SWB brochure, this illustration is very reminiscent of the Paris Salon prototype, 1539 GT. Note the lack of repeater lights, side vents, jacking holes, etc. – all would come later. At the Turin and Geneva Shows, the SWB prototype was displayed without wheels to emphasize the new disc brakes*

25

marker lights – which would be adopted later – marred its smooth sides. As befitting Ferrari's first disc-braked berlinetta, the new coupé was displayed at the later Turin and Geneva Shows with its wheels removed. Late adoption of discs notwithstanding, Ferrari left no doubt this car would go – and stop. The steering was lhd – it was later converted to rhd in the 1970s.

The berlinetta's debut was an exciting one. The car certainly looked the part and with its more powerful engine and improved braking, it promised to quickly obsolete its predecessors. Understandably, top drivers and talented amateurs once again eagerly vied for the first cars.

At Maranello, the SWB's creators included Ferrari's best engineers – Giotto Bizzarini, Carlo Chiti and Mauro Forghieri. The firm of Pininfarina again created the body design, based upon its initial efforts with the interim cars. The aggressive, yet functionally lovely

*1613 GT was the second SWB built. After its Turin Salon debut in 1959, it won the Angola GP two years running. Repeater lights and sleeved brake ducts were a later addition. Otherwise, the car retains that substantially clean look of the earliest SWBs. Photo Marcel Massini*

shape, was a natural evolution from the somewhat more angular Tour de France GTs. Scaglietti, having practiced on the TDFs and the interim cars, was more than ready to build the new SWBs.

At the introduction, word of a more refined companion model, with a steel body, sound deadening material, glass windows and a 'cooking' but still potent 3-litre engine was announced for buyers desiring a hot street car or a more comfortable club racer. This second version, called the Lusso, would be equally popular and represented a slightly more refined version of the SWB, eminently suitable for road or track work. They did not become available until *1993 GT* in July, 1960. The use of the term 'Lusso' for steel-bodied SWBs should not be confused with an SWB successor, the 250GT Berlinetta Lusso which was introduced in 1963.

It's still unclear exactly how many SWBs were built. Authorities place the total at 162. A great number of the

*Side elevation of 1741 GT shows what happened to most early cars. Window wings and fender vents were frequently retro-fitted. Note lack of rain gutters, rear window vents, license plate insert. Aside from the additions, this car is typical of the second series of early production SWBs. This car, the fourth SWB built, took a fifth at Sebring, a first at Road America and a third at Nassau in 1960. Photo Gary Schonwald*

27

Right *Trunk of 1757 GT, the fifth SWB built, shows there's little room after the cavity is stuffed with a 31.7 gallon fuel tank. Oversized competition filler pipe was standard on alloy cars. Photo Karl Dedolph*

Below *1759 GT has some interesting details – brake scoops in front of the rear wheels for cooling. This car was prepared for Le Mans with a roll bar added as well as the fender vents and a bug deflector. It finished fourth overall and seventh in GT driven by Ed Hugus and Augie Pabst*

cars have survived, fortunately, but numerous modifications and evolutionary racing changes make it a
challenge to classify them easily. From prototypes to the
later production cars, there are five fairly distinct
versions with differing design and chassis details. As
well, a number of cars were especially modified and some
were fitted with custom coachwork, during production
and afterwards.

As a general guideline, although some cars have
overlapping features, the five basic SWB variations can
be identified. A more comprehensive view of body detail
differences will be found in chapter 5.

## 1 Prototypes, 1959

The first SWBs had no side lights, rain gutters, air vents,
windwings, rear cockpit air outlets or, as yet, sleeved

1771 GT, *owned by Ado
Vallaster, is a nice example of
a reasonably early car. In the
3/4 view, it's easy to see the
side window dogleg that
would later be smoothed over.
Lack of side trim typifies
earliest SWBs. Photo A.
Martinez*

front brake cooling ducts. There were no louvres on the sail panels and the first bodies were all alloy. The rear registration plate was mounted on the decklid surface – there was not yet a recessed area to accommodate it.

## 2 Early production, 1960

Bumperettes were optional on the first production cars and many SWBs specifically destined for racing were never so equipped. An air outlet appeared in the plexiglass rear window. Brake scoops were still unsleeved and side marker lights, cooling vents, rain gutters, windwings and the recessed rear numberplate area were still to come. A few of the earliest steel-bodied cars had an aluminium trim strip below the doors.

*1811 GT is a largely original-looking SWB, except for the fender vents. The flared rear wheel wells, to accommodate larger tyres, were the mark of competition cars. Photo Alan Boe*

*Data plate on 1813 GT is found on firewall, at upper left. Car is owned by John Risch*

*Airscoops on competition cars lacked plated trim: hoods lifted off*

*Early cars lacked any cockpit venting. This second series SWB, 1813 GT, has a plexiglass rear window with the alloy scoop fastened just below the roofline*

*Doors have pronounced arch in early cars – this was later smoothed in the fourth series. Windwings here on 1813 GT were retrofitted at Scaglietti's as were the fender vents and cooling door*

### 3 Production, mid-1960

Ovoid amber side marker lights and distinctive front and rear fender vents were added for exhausting hot air from the brakes. A few cars were retrofitted with cockpit ventilation doors but this was never a production option. On most of the competition SWBs, the front brake

Left and below left *Closed, 1813's trunk is smooth; license plate insert has not yet appeared. Competition gas filler cap is in upper left-hand corner. Open, the trunk, with spare inside, leaves precious little room for anything. Not to worry, racers travel light!*

*1917 GT (R) and 2731 GT (L) contrast some of the differences between early and later cars. The earlier car, built in May, 1960, lacks windwings, the fender vents were retro-fitted, the rear window is smaller and the gas filler is in a notched portion of the decklid. The later car has a smoother window line, bigger rear window, and a gas filler in the rear fender. 2731 GT, built in July, 1961, is a lightweight SEFAC 'hot rod'. Photo Marcel Massini*

cooling ducts were sleeved and the rear wheel openings were slightly flared to accommodate larger tyres. The rear registration plate received its own recessed area and competition cars had Snap exhaust extractors. A popular addition, the Snaps began appearing on most models shortly afterward. Rain gutters appeared and would be included on all SWBs to come. Some early cars had one-piece rear bumpers – most later cars have three-piece bumpers.

### 4 Production, late 1960

The cars from this and the last period can easily be distinguished by a smoothed top window edge which eliminated the slight dogleg of the original design. As well, windwings were now standard but don't be fooled. These were often retrofitted to earlier cars. The rear

cockpit air outlet was built into the roof and the rear window was slightly enlarged for better visibility. A few of the SWBs of this period also had louvred sail panels, a throwback to the TDFs. On the sides, jacking holes were added. The spare tyre location changed from beneath the rear parcel shelf to inside the trunk.

### 5 Production, 1961–62

*Altissimo* tear-drop shaped side marker lights replaced the ovoid lights on all cars. Below the front fender vent, from *3113 GT* onward, the rectangular Pininfarina design badge was affixed, and the external gas fillers that had long been in the upper left-hand trunk corner were now moved to the left rear fender. On some of the steel-bodied Lusso SWBs, the gas filler was relocated inside the trunk. The *2111 GT* had its jacking hole located on

*2083 GT, constructed in September, 1960, represents the third series of SWBs with round marker lights, windwings, wind-up windows, fender vents, an inset license plate and a built-in vent in the roof. This competition car was originally owned by the Scuderia Serenissima racing team and, with Carlo Abate driving, won a 1960 GT race at Monza. Photo Charles and Lisa Reid*

*2111 GT is a transition car. Originally built as a street berlinetta, it remained at the works for a time and then received new panels that made it the prototype for the revised 1961 SWB coachwork. Some of the new features included a straighter side window line, Altissimo repeater lights, and a relocated and concealed glass filler cap. Car is owned by Ed Wechsler. Photo Alan Boe*

the frame. There are measurable difference in the following SWB body lines when comparing a 1960 car to a '61 or '62: cowl windshield, front and rear fender curves, grille set, shape of seatback, roof, rear window and trunk. (The author wishes to thank Alan Boe, Jess Pourret and Stan Nowak for their help and research work on SWB differences.)

Jess Pourret has pointed out that the SWBs were seldom built in numerical order. This was particularly true for some overseas deliveries. As a result, several early cars benefited from running modifications although many did not. To add to the confusion, a number of customers sent cars back to the factory for rebuilds, and modifications were often performed to update them at that time. As well, some Italian regulations were changed requiring, for example, side repeater lights, and these were frequently retrofitted on earlier cars.

Today's purist approach to Ferrari restoration, notwithstanding, SWBs were originally owned for the most part by racers who wanted their cars to be as up-to-date as possible. Thus, 'accurate' restorations are nearly impossible and almost any permutation of available modifications probably exists.

Ferrari chassis of the early 1960s were a sturdy amalgam of oval steel tubing. The SWB received its name from its shortened frame which changed from the TDF's 2600 mm (102.3 in) to 2400 mm (94.4 in.) Curiously, the 2400 mm dimension was shared with Ettore Bugatti's Grand Prix cars. Jess Pourret has also observed that a few SWBs had a longer 2420 mm (95.2 in.) wheelbase, somehow. The abbreviated chassis supposedly improved cornering speeds over the long-wheelbased variants.

The SWB's chassis differed from the TDF's in a number of additional key details. Additional tubing surrounding the upper fenders and the cowl strength-

*Here's a 1961 competition berlinetta, 2845 GT. This lightweight car has all the later features plus an outside filler cap, now in the rear fender, and flared rear wheel wells. Originally a Scuderia Serenissima racer, the car finished a creditable third in the 1961 Tour de France co-driven by Trintignant and Cavrois. Later, piloted by Jo Bonnier and Graham Hill, 2845 GT was 12th in the Montlhéry 1000 Km. Photo Marcel Massini*

*The first SWB brochure*

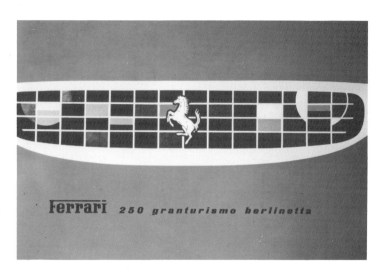

Below *Ferrari chassis were made of welded steel tubing. Extra tubing surrounded the front fenders and the cowl for additional strength. Car shown is 1741 GT, owned by Gary Schonwald, later restored by Chris Leyden*

ened the front of the car. Other new frame modifications were needed to support the fuel tank, gearbox and tubular shock absorbers. The frames and the early exhaust systems were built by long-time Ferrari supplier, Vaccari, in Modena.

The SWB's front and rear track were similar to the TDF's at 1354 mm (53.3 in.) and 1349 mm (53.1 in.), respectively. To accommodate the switch from drums to discs, stiffer springs were fitted. Adjustable tubular shock absorbers, supplied by Koni or Miletto, replaced the TDF's old lever-action type. The steel-bodied Lusso SWBs received a softer shock setting. A 15 mm anti-roll bar rounded out the front suspension on both versions. Ride height was adjustable.

In back, the rear suspension was basically a carryover

*Here is 2701 GT undergoing restoration. This alloy Comp./61 berlinetta was badly crashed at Le Mans by Jo Schlesser. Completely rebuilt a number of times, it's a beautiful restoration today. Photo Alan Boe*

*Dunlop disc brakes were the big news for SWBs. They replaced the TDF's drums with contemporary stopping power. Slow to adopt the new technology, Ferrari lagged behind Jaguar and other early proponents of the more efficient new braking systems. Tubular Miletto or Koni shocks replaced the TDF's Houdaille lever units. Photo Alan Boe*

from the LWB Tour de France cars. Semi-elliptic leaf springs, located by four tubular steel radius rods, positioned the limited-slip differential on competition cars. As well, cars destined for racing utilized polished leaf springs – ostensibly to help reduce friction. On Lusso berlinettas, the rear leaf springs, like the front coils, were marginally softer.

The new Dunlop discs were not power-assisted on the competition cars, but the front brakes did incorporate a booster pump. On the road cars, a Bendix servo made braking easier. Caliper style varied slightly – the big discs provided 490 sq. in. of swept braking area.

A newly-designed steering box from ZF, with a 17:1 or 20:1 ratio, facilitated aiming the abbreviated coupé at curves. Interestingly, the turning radius varied from one side to the other. With $1\frac{1}{3}$ turns, lock-to-lock, the turning circle to the right was 48 ft – while a full left turn required only 37 ft. The reason for this seeming contradiction was to pass the FIA scrutineering. Of course, if the drivers were required to turn right for the test. . . .

The SWB's four-speed, Porsche-patented all-synchromesh gearbox had a freshly-designed, light alloy ribbed case with new ratios. To reduce noise, the street

cars had a similar design made of cast iron, with no cooling ribs.

Exhaust systems also differed between the street cars and the competition berlinettas. Racers had only one muffler on each side and the chromed tailpipe tips were the same diameter as the exhaust pipes. Snap exhaust extractors first appeared at Le Mans in 1960, and these distinctive tailpipes were standardized shortly afterward. Street cars used double mufflers all around, from Abarth, with oversized quad tips giving a quieter, more mellow sound.

Racers came equipped with a 31.7 gallon (120 litre) aluminium fuel tank topped by a handsome quick fill Monza-type cap and an oversized inlet pipe. For the road, the steel cars had a smaller 23.7 gallon (90 litre) tank. On the early road cars, despite the quick fill cap, the inlet

*At some early showings, SWB's were displayed, as this racer is shown, without wheels – to better emphasize the new 4-wheel disc brakes. On competition cars like 2731 GT, shown here, there was no power assistance, but the front brakes had a booster pump. For street cars, a Bendix servo made stopping easier. The discs provided 490 sq. in. of swept braking area. Photo Marcel Massini*

ıs in June,
3rd, 4th
Berlinetta
reamlined
We also
mfort, so
10 as well
space for
ving easy
/ of visibi-
n be used
ifications.

| back axle ratio | 1st speed | 2nd speed | 3rd speed | 4th speed | In 4th speed per 1000 revs |
|---|---|---|---|---|---|
| 7/32 | 54 mph | 76 mph | 102 mph | 126 mph | 18 mph |
| 8/34 | 58 mph | 82 mph | 110 mph | 135 mph | 20 mph |
| 8/32 | 61 mph | 85 mph | 115 mph | 144 mph | 21 mph |
| 9/34 | 65 mph | 92 mph | 124 mph | 153 mph | 22 mph |
| 9/33 | 67 mph | 95 mph | 128 mph | 157 mph | 23 mph |
| 9/32 | 69 mph | 96 mph | 129 mph | 162 mph | 23.5 mph |

*speeds attainable
at 7000 r.p.m.
with tires size 6.00x16"*

Above *The SWB competition
berlinettas had a newly
designed light alloy gearbox
case with cooling fins. To
reduce noise, Lusso
berlinettas had a cast-iron
case which omitted the fins.
Photo of SWB brochure Dr.
Mel Wilner*

Right *SWB rear ends came
in a wide variety of ratios.
This limited slip ribbed and
finned differential is from
3269 GT, the second Bertone
special-bodied SWB. Photo
Steve Tillack*

pipe was of reduced size. This was later enlarged to racing dimensions. Filler cap placement varied from the left-hand trunk corner, with a distinctively notched decklid for clearance, to the left rear fender crown. On 1961 competition cars at Le Mans, a shield covered the left side exhaust to protect against fuel spillage onto the hot tailpipes.

SWBs were equipped with a wide variety of optional wheels for competition and street use. Initially, the cars were supplied with 6.00 × 16 (185 × 400) tyres on 42 mm Rudge three-eared knockoff hubs. Spokes were painted (and sometimes chromed) on racing cars – almost always plated on steel-bodied coupés.

Seven different rear end ratios allowed racers to select the best possible gearing for the appropriate track. They ranged from a hill climb or tight course 7/32 gearset to a high speed 9/31 ratio. At 7000 rpm with 6.00 × 16 tyres, the lowest gear topped out at 125.5 mph, while the highest gear showed a theoretical 166.1 mph capability. Pourret points out that with the SWB's blunt aerodynamics, it's unlikely that a berlinetta could actually reach this figure. Contemporary accounts list

*With a few special-bodied cars the exception, SWB's were equipped with the classic Borrani wire wheels in a wide variety of sizes. Most common at first were 6.00 × 16 tyres on 42 mm Rudge hubs. Toward the end of production, wide base 15 in. wheels were fitted*

the car's top speed in the 150–155 mph range.

For 1961, there were minor frame modifications in the competition cars, including some lighter tubing as well as additional square tubes for strengthening. Really hot competition cars received special order stiffer front springs. Most Miletto shock installations gave way to adjustable Konis – still with a different setting for racers. Polished rear springs were also continued on the racing cars.

Dunlop disc brakes remained standard on both Lusso and competition coupés – with bigger calipers finding their way onto the racers. A few cars also had their rear calipers reversed from the front to the rear of the brake rotors by their owners. Why?

The later gearboxes, still alloy for racers, cast-iron for Lussos, were redesigned and fitted with a relocated oil pump. In back, the fuel tank was repositioned in the Lussos to provide more luggage space. Conversely, the racers received an even larger 34.3 gallon (130 litre) tank, so the additional room promptly disappeared.

Fifteen inch wheels were introduced in late 1961 and a variety of wider wire wheels became available. With the smaller diameter wheels, the 7/32 rear end ratio was good for 116 mph and the theoretical top end with the 9/31 gearset was listed at 155.

Rounding out the chassis differences, many US racers were fitted with chrome-plated cockpit roll bars. Interestingly, European cars did not receive this protection.

The SWB's represented an evolutionary step toward faster Ferrari competition cars, but, as will be seen, the real efforts went into the engines so chassis development remained quite conservative.

# SWB engines

Before discussing SWB engines, it's important to note that while the majority of racing and street berlinettas fell under somewhat consistent specifications, customers, especially well-connected racers, could often opt for special treatment. This frequently included special alloy components, bigger carburettors, radical valve timing, hotter camshafts, higher compression ratios, lightweight pistons – and even varied gear ratios, suspension components, wheels and tyres. Some of this special equipment eventually became available to everyone, it's just that the privileged few got it first.

For his best customers, as we shall see, Mr Ferrari trotted out some fairly radical variations on the SWB theme. Today, when restorers do a ground up berlinetta rebuild, they can legitimately choose among a wide selection of specifications. Or, they can obtain a copy of the original construction sheet for the car – still available from Maranello if you ask nicely – and rebuild their SWB exactly as it was ordered. To begin, we'll discuss the basic SWB engines and later detail some of the special equipment.

The new competition SWBs were equipped with a 3-litre (73 × 58.8 mm – 2953 cc) V12 that Ferrari called type 168 B. Closely related to the interim berlinetta engine and the 128 DF, it was offered for about a year until September, 1960. Externally, the new engine had a horizontally-mounted oil filter, three Weber carburettors with velocity stacks, single exhaust ports and a

*Many competition SWBs used a cold air box surrounding the carburettors, fed by the hood scoop. Shown is John Risch's 1813 GT*

ribbed alloy gearbox. The Lussos had a slightly different version called the 128 F.

While the 168 B's block, crankshaft, connecting rods, pistons, camshafts, water pump and the location of the starter motor were identical to the 128 DF, updates included Testa Rossa type cylinder heads with two oil passages, coil instead of hairpin valve springs, 12-port induction and individual exhaust ports. Cam covers had 16 studs and the spark plugs were now located outside the heads.

Intake and exhaust valve sizes varied for competition and street motors. Some early competition cars received a reinforced Testa Rossa clutch. Compression ratios also varied with most alloy cars receiving 9.5:1 pistons and the steel cars using 9.3:1. A new Fispa mechanical fuel pump, mounted high up front on the left side, was an addition. As well, a Marelli electric fuel pump and an accompanying fuel filter were located, as in the TDFs and in Ferraris to follow, mounted on the lower left hand frame close to the rear wheel. The electric fuel pump was used to pressurize the system for starting – and for high speed running.

A newly designed timing chain casing allowed

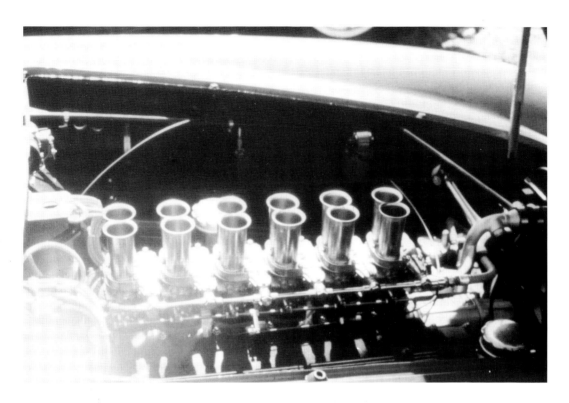

removal of the old-style 128 D/128 DF mechanical oil filter – to be replaced by a horizontally-mounted single disposable filter. As well, a second filter mounted on the inside of the right front fender, was frequently added.

The redesigned, finned oil sump had a new bolt pattern. To further save weight, some racing SWBs received special oil sumps, cam covers and timing chain covers made of elektron alloy. These cars, the famed 'SEFAC hot rods,' which will be discussed later in detail, also utilized lightweight pistons as well as specially selected and balanced, heavier connecting rods to deal with the increased compression ratios.

Street berlinetta engines, with 'heavier' pistons, milder (9 mm lift) camshaft timing and 'lower' compression ratios, developed between 220 and 240 bhp at 7000 rpm. Racing engines, good for 7500 rpm plus,

*Although no SWB was ever originally fitted with a 6-carburettor manifold, many were so equipped later. The factory listed six Weber 38 DCNs or three Weber 46s, late in 1961. Here's 2083 GT mounting the DCNs and tall velocity stacks. Photo Jerry Lynch*

*2111 GT shown here has three Weber carburettors fitted. This car was tested with Solex carburettors at the factory, later fitted with Webers. The short stacks are for maximum horsepower in a hillclimb or short track event. Street SWB engines developed between 220–240 bhp at 7000 rpm. Photo Andre Dudek*

developed 260 to 275 bhp. In typically optimistic Ferrari fashion, these motors were credited with 280 bhp. Still, it's likely that the special equipment engines built for the SEFAC cars did approach 280–290 bhp.

Racers and some street cars alike used three 40 DCZ/DCL 6 Webers with velocity stacks and a cold air ram induction aluminium air box fitted flush under the hood scoop. Velocity stack length varied as to whether the car was slated for short (hillclimbs or sprints) or long distance work.

Many street SWBs came equipped with a standard, flat or gloss black finished air cleaner – a more practical application than the open intake stacks for everyday driving. Solex C 40 PAA L carburettors were apparently tried on a few SWBs, too. Homologation papers filed for the cars ensured Ferrari would be able to capitalize on some fairly exotic engine modifications. Dry sumps were registered by March 1961, and in July of that year, Ferrari listed six Weber 38 DCNs or three huge Weber 46s as possible variations for competition SWBs. Alan Boe believes this was done as much for the forthcoming GTO as for the GTB.

In keeping with the numerous variations on street and racing SWBs, the exhaust systems were no exception. Most racers had a pair of six branch steel headers joined in a single wide collector pipe. From 1961 onwards, many racers were fitted with tuned split manifolds, divided into two bunches of three exhaust headers each. A detachable aluminium heat shield was available for both types – lined with asbestos to further absorb the exhaust heat and to help keep the engine from roasting the cockpit's occupants.

Up to the end of 1962, SWB engines had two oil breathers/fillers located between the exhaust manifolds, one on each side of the engine. Later, these were moved to the front of the engine where they remained through the later 275GTB series.

SWB ignition systems were fairly standard with two, twin-point vertical Marelli distributors, driven from the camshaft ends. Dual Marelli high-voltage coils were firewall-mounted. A 300-watt offset generator, first seen mounted centrally on *1523 GT*, the last LWB interim berlinetta, would be standard equipment on SWBs and would continue through the 250GTOs.

For cooling, although some racers omitted the fan, alloy and steel-bodied cars shared a 15-litre, thermo-statically-controlled newly designed copper radiator and a cockpit-operated radiator blind. To round out the differences, racers featured a single Marchal horn – Lusso street cars had twin horns activated by compressed air.

Ferrari authority, Jess Pourret, points out that, by the fall of 1960, the Ferrari factory's production of 250GTs included the two SWB variants, two California Spyders (one with a racing engine, the other with a detuned street motor), the 2+2 250GTE and the Series II

*Some street berlinettas, like 2177 GT, had competition Weber carburettors and this car had an alloy finned gearbox fitted as well. This sort of thing gives restorers nightmares! Brake servo is shown to the right of the engine. Photo Alan Boe*

Pininfarina cabriolets. Consequently, to simplify engine production for the limited quantities of each of these cars, the factory standardized some engine improvements developed during the previous season's racing.

The type 168 engine, introduced at this time, had a new silumin block typed the 128 E. This new second version motor had four studs per cylinder to eliminate compression leakage and a third oil passage was added for better cooling and lubrication. SWB Lussos and racers, the cabriolet, both California models and the 2+2s would utilize this basic block. For 1963, with some additional modifications, the 250GT Lusso berlinetta and the 250GTO would continue with it, too.

Most of the bolt-on components remained constant, although lightweight flywheels were supplied for some racers. Late in 1960, the cylinder heads were redesigned

to accommodate the aforementioned four studs per cylinder and the additional oil passage. Cam covers now had 14 studs (versus 16 previously) and the exhaust ports were changed. Now, the header pipes were separated and the two pairs of inside pipes were joined – but not siamesed.

Competition cars received bigger valves as well as bronze valve seats and valve guides – yet another example of the careful refinement that characterized the differences between the two visually very similar models.

At the front of the engine, the timing chain cover was reshaped to accommodate two vertical oil filters – a practice that would continue on Ferraris through the range of V12 engines to come. Oil filters/breathers were relocated to the front of the block at this time.

*Most street berlinettas, like 4051 GT, shown here, had conventional air cleaners in a design that Ferrari used for many years. For street driving, this is a more practical set up. Photo Alan Boe*

*When the 168 B block was introduced for SWBs, a newly designed timing chain casing facilitated removal of the old-style mechanical oil filter. Its replacement was, at first, a horizontally-mounted single oil filter. Later engines, like 2111 GT, shown here, had a vertical filter and a second filter was fitted, mounted on the front fender. Photo Andre Dudek*

For some SWB competition cars, the real 'hot setup' for 1961 was a very special, limited production engine called the type 168 Comp./61. These engines were built from February to July of that year. As well, at least one of these engines was supplied to a competition SWB in 1962. The 'normal' competition motor was infinitely more tractable than this modified version. Reportedly, camshaft timing was revised to provide an urgent, some say brutal, burst of power at 5500 rpm. Below that piston speed, the overcarburetted and wildly timed engine lacked flexibility. The later 250GTO engine, with revised timing yet again, would smooth out its available additional power throughout an even higher rev range. Before that could happen, racers simply had to learn to deal with a Comp./61's savage mood changes.

Popularly known as 'SEFAC hot rods,' a special series of 18 ultra lightweight cars were built in 1961 to utilize

*A Fispa mechanical fuel pump, mounted low on the left side, ensured a constant supply of gasoline. For starting and high speed driving, the Marelli electric pump, frame-mounted on the lower left rear of the car, pressurized the fuel system. Photo Andre Dudek*

*Twin Marelli high-output coils and vertical, dual point distributors handled the sparking chores. Photo Andre Dudek*

the Comp./61 motor's extra power. Ferrari was determined to win the Constructor's Championship that year and these variations on the SWB theme were built to make sure that happened. In the hands of talented privateers, these cars won their share of races. They include, less the *GT* suffix, the following cars: *2417, 2439, 2445, 2687, 2689, 2701, 2725, 2729, 2731, 2733, 2735, 2767, 2787, 2819, 2845, 2937, 2939, 2973,* and *3005.*

As well, and mentioned in the chapter on special-bodied SWBs, two Pininfarina SA-bodied cars, *2429 GT* and *2643 GT*, also received these engines. Two other cars, *3327 GT* and *3539 GT*, were reputed to have Comp./61 engines but SWB expert Alan Boe and I believe this is unlikely. *3327* was fitted with same, but not all competition features, and as *3539* may have been used for some GTO studies – but, at the time of writing, its history is uncertain. Since the GTOs were active when it was built, it's unlikely the car had a Comp./61 motor.

Predictably, Ferrari swept the Championship in 1961, far ahead of runner-up Porsche. Today, the SEFAC Comp./61s are some of the rarest and highly valued SWBs – if you can find one of these today, you've got pretty close to the ultimate SWB.

Comp./61s had Elektron alloy sumps and timing covers, a single oil filter mounted vertically on the front right-hand side, Testa Rossa style cylinder heads with bigger valves, a 10 mm high-lift camshaft 9.7:1

compression, lightweight pistons, polished connecting rods, split-tuned exhaust headers and a bigger od exhaust system. To cap off the improved breathing, the cars featured tureen-sized Weber 46 DCF 3s – big enough for a 4-litre car.

The oversized carburettors, while appropriate for high speed work, make street driving today very difficult, so restorers are well-advised to fit smaller DCZ 6s unless they plan to use their cars exclusively for historic racing.

The 1961 cars continued using single mufflers on each side and Snap tailpipe extensions. As mentioned, some special cars had oversized exhaust systems. Street berlinettas continued their use of Abarth systems with four mufflers and oversized, slanted exhaust tips.

For 1962, most of the 35 SWBs built were street cars with steel bodies and milder engines – with the aforementioned single exception.

A complete list of engine variations will be found in the appendix. It's easy to see that, having developed the basic successful engine design at the SWB's inception, the wizards at Maranello continually refined these motors, using all the known techniques to improve gas flow, aid engine breathing, ensure rev increases and strengthen the powerplants for longer racing life. Suffice it to say, SWB engines, from mild to wild, ensured the cars stayed competitive in any environment.

Enzo Ferrari's emphasis was always on engines. As he stated in his memoirs, *My Terrible Joys*, published in 1963, 'I have always given great importance to the engine and much less to the chassis, endeavouring to squeeze out as much power as possible in the conviction that it is engine power which is – not 50 per cent but 80 per cent – responsible for success on the track.'

Ferrari was eventually to change this view and place increased future emphasis on chassis design, but, as will be seen shortly, the SWBs were built in an era where chassis development at Ferrari was extremely conservative and experimentation definitely took a back seat to high horsepower.

# Coachwork

SWB coachwork, with special-bodied cars the exception, of course, can be divided into five fairly distinct body types. Modifications and 'repairs' through the years, as well as the factory's tendency to 'hold' overseas orders at times, have blurred the distinctions a little but basically, the cars were built as follows:

The first two prototypes, *1539 GT* and *1613 GT*, constructed in 1959, did not have the distinctive vertical fender vents, front and rear, nor did they have side signal repeater lights. The indentation for the rear registration plate, on the decklid, had yet to appear – the same was true for the side window rain gutters. Sliding windows would also be a later development, on competition cars and the front brake inlets, at first, were simply holes cut in the lower panels – these would later be finished with oval-shaped sleeved ducts.

The side windows were not fitted with vent windows. As well, the early cars had a slightly 'notched' side window line. Lightweight bumpers appeared on the first prototype.

With the exception of the third SWB chassis, *1739 GT*, which sported a custom Bertone body described fully in chapter 6, the next 11 cars had standard SWB bodies. In the early months of 1960, bumpers were optional – and a few cars did not receive them. By 1961 an air outlet appeared at the top of the plexiglass or glass rear window. All of the next 11 SWB-bodied cars (from *1741 GT* to *1849 GT*) were alloy – and all had left-hand drive.

The remainder of this second series of cars' characteristics were basically identical to the prototypes'.

The third series of SWBs, starting with *1875 GT* – and running from May to August, 1960, featured the vertical SWB fender vents, front and rear, as well as ovoid repeater lights in the front fenders. Other differences, from the rather plain early cars, after *1917 GT* included a sculpted insert for the rear registration plate.

Sliding or wind-up windows were available, still without vent windows. On the top edge of the rear window, a small vent was introduced for cockpit cooling. Brake cooling ducts were now sleeved on these cars. The first right-hand drive SWBs commenced with *1993 GT* and that car was the first Lusso, steel-bodied SWB.

Finally, small rectangular red reflectors were added to the inside of each taillight cluster next to the lower

*The earliest cars lacked fender vents, repeater lights, a license plate indentation and a cockpit rear roof vent. Cars like 1771 GT (right) were cleanly simplistic. Photo Marcel Massini*

corners of the trunk.

The fourth series of substantive changes took place from August to December, 1960. The SWB's cockpit air outlet was moved from the rear window itself to become a cleanly drawn roof vent. The rear window was enlarged for greater visibility. Lusso cars, with windup side windows, were fitted with vent wings. Rain gutters appeared, parallelling the newly reshaped side window lines beginning with *2399 GT*. On a few SWBs, louvred sail panels, reminiscent of the TDFs, were found, but this feature was not to be continued.

As well, these 'street' berlinettas were fitted with the big Monza-type filler cap leading to a small inlet pipe and a smaller tank than the racers. Jacking holes, with chromed plugged covers, appeared on each side. Lastly, competition cars had flared rear fenders to accommodate

*At first, side windows were not fitted with vent wings. The earlier cars, like this rhd example, had a slightly 'notched' side window line. On competition cars, front brake inlets at first were simply holes – later they were sleeved as this example shows. Side repeater lights initially were round, later these gave way to Altissimo oval lights. Photo David Seielstad*

a wide range of larger racing rubber.

The fifth series of SWBs, produced from early 1961, through the end of the following year, beginning with *2399 GT*, are easily distinguished from their predecessors. Besides the altered window line – which more closely matches the curve of the rear fender – they can be identified by the relocated gas filler – which migrated from the left-hand trunk corner to the left rear fender. Some Lusso cars had the filler inside the trunk. To make things even more confusing, reportedly, *3327 GT* has a

*A frequent occurrence, this competition SWB, 1813 GT was retro-fitted at Scaglietti with Altissimo lights, fender vents and an air inlet door to cool the cockpit. Car owned by John Risch*

*SWB at speed, driver adopting contemporary straight arm driving stance. Car looks slightly 'up' at the front*

non-factory fuel inlet on the right rear fender, but this was an unusual exception.

This last series had the largest number of cars sharing basically common design features. Teardrop Altissimo running lights replaced the rounded repeaters of the previous series. Finally, the jacking points were re-located into open ended chassis tubes. On competition cars, quick lift jacking points were used front and rear.

Throughout the SWB's run, changes in roof vents, the location of the oil pressure gauge and the placement of the spare tyre varied. SWB historian Alan Boe says that '. . . these developments don't necessarily correspond exactly with the changes from one body style to another.' It's maddening to historians, but the five body variations simply don't follow in strict chronological order. Borrowing from Jess Pourret's and Alan Boe's excellent research, here are just a few examples of the confusion. *2237 GT*, built in November 1960, is a fourth series alloy car and it should be the last of its type. However, *2291*

Left *It's hard to find an early car that hasn't had fender vents installed. Here's 1917 GT, owned by Sigi Brunn. Photo Marcel Massini*

Below *Later competition cars such as 2701 GT, second from left, had filler cap moved to the left rear fender, full cockpit vent in roof and license plate insert. This car also has competition identification lights and a shield over the left-hand exhaust pipes to prevent a fire from fuel leakage. Rear fender openings were flared to accommodate larger tyres. Photo Alan Boe*

2111 GT, *the transitional Lusso berlinetta that became the model for the 1961–62 last series of cars, shows off its comfortable interior and leather bucket seats. Rain gutters are featured on street cars now and a revised curve in the side window line make these later cars a bit more attractive than the earlier models. Photo Alan Boe*

GT, a steel car, and *2321 GT*, an alloy body built late in 1960, are similarly bodied, overlapping the final SWB 1961–62 body variation. In the middle of these is *2177 GT*, built in late October, which has the upper design features of *2111 GT*, the last series prototype. Suffice it to say, there's not a lot of logic here and while you can try to separate SWBs with predominantly similar features, the exceptions can drive you crazy!

In general, competition cars had glass windshields and all the other windows were plexiglass. The Lusso SWBs had all glass windows. Up front, the grille size also varied. Initially, the prototype's squared intake featured a prancing horse and twin Marchal fog lamps. At times, later cars omitted the *cavallino rampante* and other types and sizes of fog lights were fitted.

In typical Ferrari practice, the bumperettes and the rear bumper were all but useless as they were attached to the bodywork, not the frame. There are minor differences in the bumperettes from 1960 to 1961, the latter being a little smaller (and even less useful) than their predecessors.

Both competition and Lusso berlinettas had a full alloy belly pan, sometimes made in two parts, often in three. The pan ran from just behind the front of the grille and ended at the edge of the gasoline tank.

Hoods on SWBs also exhibited varied treatments. Some had louvres cut alongside the cold air scoop. The scoop itself, on Lusso cars, had a chrome trim ring. Again, a few SWBs had leather straps to retain the hood in place – others used the characteristic chromed Ferrari

*On the way to the final SWB configuration, here's a late competition berlinetta on a misty Italian morning, ready to spring into action. Photo the late Peter Coltrin, courtesy Dean Batchelor*

*Grille sizes and lighting vary. Here's 3337 GT, a 1962 Lusso. The car's large lights nearly fill the grille. Even with bumpers fitted, these cars are very aggressive looking. Photo Dean Batchelor*

hooks.

The bodies of the cars are made up of many small pieces of metal, shaped and welded together. An SWB roof, for example, is made up of over a dozen pieces welded together.

Toward the close of 1960, as mentioned, a few steel-bodied cars had louvred sail panels similar to those on *2283 GT*. For the final series, the windshield was enlarged slightly and the roof section flattened. It's a subtle difference, but it's evident when two dissimilar cars are parked alongside one another. As well, at this time, the rear section was widened slightly to accommodate larger tyres.

Other detail differences of interest to restorers are the fact that competition car trim is aluminium, while Lusso trim is chromed steel. Competition cars have understandably minimal soundproofing and no trunk carpet-

'Splendor in the Grass' -
1917 GT, *currently owned by
Ernst Schuster, is a
competition SWB with a brief
but successful racing history.
The car's best finish was in
1962, it won the GT class in
the 1000 km Nurburgring
race averaging 123 km/h.
Characteristic of early cars,
there are no vent windows
and the rear side repeater lights
were omitted. Photo Marcel
Massini*

A colourful pair - here's 1807 GT, a competition SWB shown in contemporary 1960 photo at the Nurburgring with a Fiat Abarth 750. This car's record includes a 2nd OA at the Montlhéry GP driven by W. Seidel in 1960, and a 1st OA at the 500 km of Spa co-driven by Seidel and Willy Mairesse. Vent windows were retrofitted in 1961, along with an engine from a Testa Rossa. Photo Karl Dedolph

*A beautifully restored Comp./61 engine in 2701 GT. A Scuderia Ferrari racer, used to test the new hot motor, 2701 was later crashed by Jo Schlesser at Le Mans. The short velocity stacks were suitable for hillclimbs and, in fact, this SWB competed in some Austrian hill events. Rebuilt and restored, it's a stunning example today. Photo Alan Boe*

*'Run what ya brung' An unidentified very early SWB at Thompson, CT, in 1961 shows that taped headlights, hastily added numbers and a change of plugs were all that was necessary to transform a tractable street car into a competitive racer. A close look inside shows the chrome-plated rollover bar fitted to cars bound for the States. The lack of cooling vents is typical for the earlier SWBs. Photo Kurt Miska*

Left 2335 GT *is a steel-bodied SWB built early in 1961. Featuring a built-in roof vent, wind wings, outside gas filler, 40 DCL Webers and steel hood and deck, this car is typical of the fourth series production. According to Jess Pourret, the berlinetta was found lacking in horsepower, when delivered, but a trip to C. Pozzi's in Lavallois soon restored its punch. Owned by Henry Chambon, it was photographed at a visit to Modena in 1983. Photo Alan Boe*

Above Left 2111 GT *is a steel bodied SWB that was used as the prototype for the 1961 bodywork. The hood and deck are aluminium. A neat cover in the left rear fender conceals the gas filler cap. Note Altissimo teardrop repeater lights, no side trim or jacking holes as found on the later cars. This car was tested with an electric overdrive and Solex carburettors. Photo Alan Boe*

3269 GT *was Bertone's second styling effort on an SWB chassis. Beautifully restored by Steve Tillack, and shown at Pebble Beach, this stunning car is often thought to be the most attractive special-bodied SWB. Enzo Ferrari was reportedly miffed when Nuccio Bertone sold this car; it had been given to the coachbuilder with the express purpose that it not be sold. Unique ovoid vents and gentle curves are a contrast to standard SWB styling*

3175 GT *was used for GTO studies. For a while, it was equipped with 46 DCZ 3 oversized Webers. After the GTO development work, it was rebuilt as a normal steel-bodied SWB. It was raced at Daytona in 1964, finishing 12th OA. Owned today by Giuseppe Lucchini. Photo Marcel Massini*

Top 3477 GT *is a steel-bodied car with an interesting history. The present owner, Herr Fehlman, found the car in 1974 in Germany missing its engine, drivetrain, dashboard and other interior components. As well, the left front fender was gone and the*

*frame had been butchered. The resulting restoration saw the car rebuilt to a very high standard - a fitting tribute to SWBs and an acknowledgement that the relative rarity of these cars makes such a restoration a paying proposition. Photo Marcel Massini*

Above *Here's a line-up for you - four SWBs and a Daytona! From back to front there's 2701 GT - a Comp./61 racer, and three street Lussos - 3359 GT, 3657 GT, and 4065 GT. The latter was the last SWB built, delivered in February, 1963*

*4065 GT, the final SWB, was a steel-bodied car with a very basic Lusso engine. The car was, however, fitted with 20:1 steering (like the succeeding 250GT Berlinetta Lusso) instead of the 17:1 ratio common to SWBs. The rear view emphasizes the car's sleek lines and, on this late version, the gas filler has been moved inside the trunk. Photo Alan Boe*

*Owned by Gerry Gamez here's 3035 GT competing at the 1984 Monterey International Ferrari Meet. 3035 is a Lusso steel-bodied berlinetta. The side exhausts are unusual - they make quite a sound at speed. Photo Bob Dunsmore*

Above *The long-wheelbase 250GT Berlinetta Tour de France of 1958-1960 was the direct ancestor of the 1960 250GT SWB. As ever, this car was bodied by Scaglietti from a Farina style. This car's wheelbase was 200mm/7.9in. longer. Even so, until the 250GT SWB came along in 1960, this Ferrari was still favourite to win any GT race in which it was entered, while road-equipped versions were definitely the fastest cars on the highway*

Left *The number plate on this 250GT SWB complements the sensational styling, in which not a line, not a cupful of space is wasted. Although theoretically such cars could be used on the road, the competition versions carried no bumpers, while the panels themselves were very thin. Slots behind the rear wheel-arches were to help extract hot air from the brakes, and on this example the petrol filler cap is covered by a flap*

Above *A famous car when it was new, and now equally famous in later life, is this blue SWB Lusso race car. The colour scheme, complete with white band across the nose, tells its own story, for this was the second of two different 250GT SWBs in which Stirling Moss won the Tourist Trophy race at Goodwood, UK – in 1960 and 1961. Now carrying a special (and very appropriate) British registration plate, it originally carried an Italian identity*

Right *Suitably modified for occasional road use in London (as here), the ex-Tourist Trophy winning SWB Lusso race car is exhilarating transport. So what if there is no protection from heavy traffic scrapes, and so what if interior noise levels are high? This is a legendary machine!*

Left *The important instruments were ahead of the driver's eyes, with a complementary array of auxiliaries in the centre of the facia. As on all such Ferraris, the gear lever knob was set up high, very close to the (wooden) steering wheel rim – the 1962 250GTO would be the same*

Below *Comp./61 versions of the Colombo-designed 2,953cc V12 used three downdraught dual-choke Webers, with open trumpets. Minimum peak power was 280bhp@7,000rpm*

This silver-painted 250GT SWB was the winner of the 1960 Tour de France, when it was driven by Willy Mairesse, with co-driver G. Berger. This car, carrying chassis number 2129GT, was followed home by two other 250GT SWBs, totally dominating the gruelling ten-day event. Forty years after its famous victory, 2129GT still carries the competition number of 157 with which it won the event, together with the chequered band above the radiator inlet. After Mairesse had finished with it in the 1960 Tour de France, however, the front end was rather crumpled. To quote The Autocar's race report it '... came into contact with many solid objects' – some of which were other competing cars. The view below right shows off the four Snap exhausts and neatly finished rear brake hot air cooling outlets

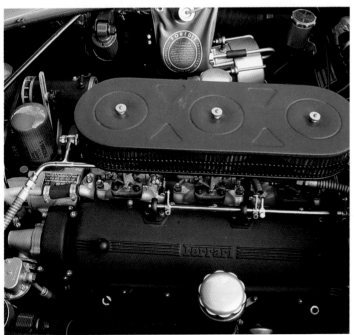

Above This beautiful example
was one of the very last
250GT SWBs, originally
delivered to Lord Portman in
1962, equipped as a road car.
This view shows the neat
(but rather ineffective?) rear
bumper blades, and the final
type of flap for the filler cap

Left When 250GT SWBs
were sold for road use, they
were usually equipped with
this massive air-cleaner over
the three downdraught carbs
with six vertically positioned
trumpets. To Ferrari buffs,
this was a familiar
installation, as all earlier
models had used the same
V12 and this air-cleaner had
already appeared on several
other models. Even in 'detuned' road car form, a 250GT
SWB had at least 240bhp and
top speed approaching 150mph

Left *There was really no such thing as a 'standard' 250GT SWB. In comparison with the ex-Moss/ex-Rob Walker race car of 1961, the steering wheel is dished instead of flat, there is perhaps a little more gloss on the panel itself, and there are enough details to spell out 'road' instead of 'race' car.*

Below *Road-equipped 250GT LWB machines usually carried slightly heavier bodywork (some of the panels had thicker gauge metal too), and quarter bumpers were available to fend off other traffic. Later cars, like this 1962 example, also had 'tear-drop' shaped indicator-repeater lamps alongside the headlamps*

**Right** *Not a curve out of place, and not a superfluous fitting in sight. Like all other road-equipped 250GT LWBs, this metallic blue example had a separate boot compartment (which was mostly full of a vast fuel tank and a wide-rimmed spare wheel!)*

**Below** *Seating for two, but little other space to spare in a 250GT SWB. Who needs more? An SWB owner could carry very little luggage inside the cabin with him; there was little space in the 'luggage compartment' either – so it was as well to travel light, or to have a separate tender car. Not that Ferrari cared about this, as the SWB was essentially a competition car which a few, fortunate customers wanted to use on the road*

ing. Lussos had, befittingly, more luxurious interior trim with leather seats and leather-topped dashboards. Over two dozen cellulose colour combinations could be specified although *rosso chiaro* seems to have been the predominant favourite.

Lusso cars had steel bodies and most of these had alloy doors, hoods and decks. Alloy SWBs were constructed at Sergio Scaglietti's of aluminium. For the SEFAC cars, a paper thin alloy was fitted that was so light it would dent if you looked at it sideways.

Le Mans were modded with a rear left side exhaust pipe guard to prevent fuel spillage after the tank was filled. As well, these cars received coloured marker lights on the roof and doors for night race identification. Bug deflectors on the driver's side, and aluminium fender flaps to prevent rock kickup, were also found. A few racing berlinettas even had extra mini scoops for rear brake cooling.

The SWB's cockpit evidenced some evolutionary differences from the TDFs. On early competition cars, the dash panel, finished in black crackle, had a big 8000 rpm tachometer and speedometer nestled alongside one another. Speedometers were marked in mph or km/h,

*On early competition cars,
like 1813 GT, the spartan
dash was finished in black
crackle and the speedometer
and tachometer fronted the
driver, with the balance of the
gauges off to the right*

*On later street cars, like 2111
GT, the dash was leather
covered and the lower panel
was painted to match the
exterior. The oil pressure
gauge left its mates to reside
between the speedo and tach
dials. Photo Andre Dudek*

depending upon country of intended use.

To the right of the twin dials, in the earlier cars, the oil pressure and the oil temperature gauges, water temperature, gas gauge and the clock can be found. Glove compartments were omitted on competition cars, but they were included on street cars.

Ventilation was never really adequate on these tight little coupés. Three push-pull levers on the left side of the wheel activated the left and right air inlet pipes as well as the hot water tap for the heater. The radiator had a roll-up blind controlled by an alloy wheel to further regulate radiator air intake and bring the engine quickly to operating temperature.

Rounding out the acoutrements, a proper dead pedal to the left of the clutch, provided a resting place for the driver's left foot. The fly-off handbrake was conveniently (in the left-hand drive cars) placed just to the left of the transmission tunnel.

The remainder of the interior fittings were similar to

*Later competition cars, such as 2701 GT, still had the crackle finish, and their oil pressure gauges migrated over as well. Photo Alan Boe*

*Radios had to be an afterthought in a cramped SWB cockpit. The owner of 3337 GT fitted a radio atop the transmission tunnel. Photo Dean Batchelor*

most street Ferraris – with door opening and courtesy lights, a dipping mirror and an ash tray. Again, in typical Maranello practice, the reclining back seats in a few cars had a headrest just for the passenger. Presumably, the driver was considered to be too busy for such niceties.

On competition cars, the floor pan was a single sheet of aluminium. This was doubled on steel cars and more soundproofing materials were added. The lighter cars had perspex sliding windows. Some slid vertically, some horizontally. Thus, with the winding mechanism

omitted, a deep box in each door provided limited storage space. On Lussos, door pockets served this purpose.

Upholstery was leather or leatherette, as specified – and carpeting was dyed to match. For 1961 and 1962, the interior changed slightly. On competition cars, sliding perspex windows gave way to pull-up plexiglass with a hold-down strap to fix the windows in any of seven positions.

On the dashboard, the twin tachometer and speedo-meter dials were separated slightly and the all-important oil pressure gauge was placed between them. A single curved cover now surrounded the three dials. Com-petition cars still largely used crackle finish, while on the street cars, the vertical portion of the panel was usually painted to match the car's exterior.

For the final years, the early SWB's dark-faced

*A nice example of a later street car, 3233 GT, built in February, 1962, lacks bumpers and has an outside gas filler for a competition look. Outside jacking holes have appeared and the repeater lights are now Altissimo ovals. Car owned by Gary Schaevitz. Photo Alan Boe*

Right and below *One of the best known Comp./61 cars is 2689 GT. In 1961, driven by Noblet and Guichet, it finished third OA at Le Mans, first in GT category. It has a distinctive pit marker light on its right rear fender. Photo Andre Dudek*

*Closeup of the exhaust shield on 2689 GT. Photo Andre Dudek*

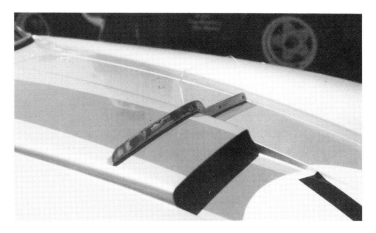

*Close up of bug screen on 2689 GT – it's marginally different from the unit on 2701 GT. Photo Andre Dudek*

instruments were replaced with grey faces. Toggle switch location varied as did the location of cigarette lighters and ashtrays.

Leather or leatherette seats, sometimes with cloth inserts, could be specified. The carpeting still matched the main upholstery colour in most cases, although special colours could and did provide exceptions. The traditional Ferrari wood rim steering wheel was slightly reduced in diameter from 420 mm (16.5 in.) to 400 mm (15.7 in.) for better steering control.

For purists, chassis numbers of SWBs can be found in a number of places besides on the official data plate, the

*Closeup of rear fender vent and splash guard on 2689 GT. Photo Andre Dudek*

*Sleeved brake scoops on each side of 2689 GT directed air to the disc brakes. Photo Andre Dudek*

engine block above the starter motor and the front top centre of the engine. The number may also be found stamped on the hidden surfaces of a number of components such as the grille, door sills, glove compartment door, hood and interior window frames.

Serial numbers have also been found written inside the dashboards, scrawled under seats, on the firewall

and in other locations. As these components were assembled separately at the factory and at Scaglietti's, it's logical to assume that the numbers served as factory identification during assembly. Interestingly, gearbox numbers differ from chassis numbers although they follow approximately the sequence in which SWBs were assembled.

Perhaps the best way to authenticate any SWB is to obtain, from the factory, a set of the car's original assembly data sheets detailing all standard and optional equipment as the SWB left Maranello. Of course, these don't show subsequent modifications, so they can be misleading about a car's present appearance. Still, they are a helpful guide to determining original equipment.

For restorers, it's hard to blame someone for changing, within a contemporary Ferrari context, the equipment on an SWB restoration – fitting leather seats, for example, where leatherette was originally supplied, effecting some Comp./61 engine mods, installing side vents, vent windows, bigger wheels and tyres, etc. – all fall prey to this registration upgrading.

I believe it's best to restore the car to its original condition, but if leather seats and a little more power are your wish, I'll not argue the point. Suffice it to say, a well-restored SWB is a delight no matter how it was equipped.

*4065 GT, the last SWB, shows how the features evolved. The gas filler cap on this Lusso is covered by a fender flap, and the SWB's smooth lines are unbroken by flares, flaps and added lighting. Photo Alan Boe*

73

# Special *carrozzerie*

Although the 250GT SWB was primarily designed with one basic body, it didn't take long before some experiments with alternative coachwork appeared. Relatively few special-bodied SWBs were built, but the ones extant are interesting from both a styling and a development point of view.

Two Bertone bodies, *1739 GT* and *3269 GT*, differ greatly from standard SWB styling. *1739 GT* was the first custom-bodied SWB, built on the third chassis, and it was completed early in 1960. Designed by Giovanni Michelotti, the car features crisp, still contemporary lines. At various times, it appeared with and without fender repeater lights, and with at least three different colour combinations – a light green body, a dark roof with a light-coloured body and a dark body with a light roof.

Snap exhausts were fitted and the car featured slotted Campagnolo alloy wheels – sometimes with two and other times with three-eared knock-offs. After passing through a few owners in Italy, the car came to the US and was recently restored by Steve Tillack in Carson, CA. The dashboard of this car anticipated the 1963 Ferrari Lusso in design. Distinctively un-Ferrari-like in conception, *1739 GT* is nevertheless a very attractive variant. Looking at this coupé today, alongside a sister standard SWB, it's evident Bertone's efforts struck a timeless balance – the original Pininfarina SWB is definitely dated, whereas this special car seems to gracefully transcend its era.

*1739 GT, the first Bertone-bodied SWB, was the third car built. It's very different from its competition-oriented brethren, with modern, very contemporary lines and a unique interior treatment. Campagnolo disc wheels were an unusual feature as SWBs were customarily equipped with Borrani wires. Photo Steve Tillack*

*Viewed from the rear, 1739 GT has truncated tail, revives some SWB feeling with its vertical round taillights. Snap exhausts subtly emphasize the car's competitive essence. Photo Steve Tillack*

*Beautifully restored engine compartment of 1739 GT has velocity stacks for a competition flavor. Steve Tillack did this exquisite restoration from what was a badly deteriorated car. Photo Steve Tillack*

*Dashboard in 1739 GT anticipated the later 250 GT Lusso with its centred speedometer and tachometer. Unique seats, with deep padded rolls, represent the height of luxury for a grand touring car at the time. Photo Steve Tillack*

Next, chronologically, *2429 GT* was built in May, 1961, by Pininfarina. An all-alloy body, *2429 GT* was designed to resemble the 400 Superamerica enclosed headlamp coupés. The car was built to be as light as possible. Jess Pourret reports the chassis was drilled for additional weight reduction.

Under the hood, this featherweight GT benefited from

*2491 GT, a striking cabriolet currently owned by Anatoly Arutunoff, is a Zagato-bodied SWB that was first exhibited at the 1974 Turin Show. It's possible that a second similarly-bodied SWB exists as well. Owner Arutunoff has been trying to research this car's history for some time. Slotted headlight grilles and flared fenders are just a few of this car's distinctive features. Photo Anatoly Arutunoff*

a full bore Comp./61 engine with oversized valves, high lift cams, lightweight pistons, big 46 DCF Webers, a competition exhaust system and a special aluminium fuel tank. Reportedly, the modified engine developed 285 bhp and, while the car was never raced, it was reputed to be very quick. Pourret reports this unique SWB is still in France, in well-preserved condition. Altogether four Pininfarina SA designs – incorporating various differences – would ultimately be built on SWB chassis.

A very similar car, *2643 GT*, is generally considered to be the prototype for the 250GTO. Built by Pininfarina, once again along Superamerica lines, it also came equipped with a Comp./61 engine, this time with dry sump lubrication. Alternatively, the engine was tried with six Weber 38 DCNs and three oversized Weber 46s. *2643 GT* competed as SEFAC's Le Mans entry in the experimental class and was later sold to Luigi Chinetti whose North American Racing Team campaigned the car at Sebring and other major events.

Anatoly Arutunoff owns the next custom-bodied SWB, *2491 GT*. This attractive car, on a Lusso SWB chassis, wears a cabriolet body, called the 3Z, styled by Zagato. Its very contemporary lines are unlike any of the special SWB coachwork – and, in fact, the car really doesn't resemble any other Ferrari custom bodies. A lovely machine, finished in dark blue, it attracts attention

*2821 GT was the fourth of five different Superamerica-bodied SWBs. This car, exhibited at the 1961 London Show, has an open headlight front end design that's reminiscent of the 250 GTE styling. Other Superamerica variants had enclosed lights and some had full Comp./61 engines. Photo Karl Dedolph*

whenever it makes an appearance.

Arutunoff believes his car '. . . is one of at least two, and possibly three such Zagato designs on Ferrari chassis.' A second, somewhat similar car is illustrated in Michele Marchiano's book, *Zagato*, and the illustration of Arutunoff's *speciale* in that book shows it with slotted cast magnesium wheels versus the Borrani wires it currently wears. The *2491 GT* started out as a 1961 steel-bodied Lusso and was later converted by Zagato at Luigi Chinetti's order. Exhibited at the 1974 Turin Show, it was brought to the US sometime afterwards and sold by Luigi Chinetti. Arutunoff acquired it in 1980 and has been trying to trace its history ever since.

The third Pininfarina SWB *speciale* was *2613 GT*. This one-off resembled *2429* except that the front end design had open headlights and a grille very reminscent of the contemporary 250GTE. Built for a member of the Dutch Court, the car also differed from the previous Super-america SWB variants in that it was fitted with a 'normal' Lusso engine.

*Bertone's second and final design on the SWB chassis was 3269 GT – an exquisite car many feel is the most attractive SWB design variant. Meticulously restored by Steve Tillack, who also restored 1739 GT, the first Bertone car, the curvaceous berlinetta is a very valuable car today. Photo Steve Tillack*

The fourth and last SWB with 400 Superamerica-type coachwork was *3615 GT*. This car closely resembled the open-headlight *2613 GT* and was initially built to special order by an Italian client, F. Gatta. This attractive car was recently offered for sale in the US. It's equipped with a standard SWB engine.

Bertone's second and last styling effort on the SWB chassis was *3269 GT*, a truly beautiful one-off with a split nostril nose reminiscent of the Ferrari GP racers of that era. This curvaceous berlinetta, on a Lusso chassis, had unique ovoid engine vents. Running gear included Borrani wire wheels and an Abarth exhaust system.

This second Bertone car was the Geneva Show model of 1962 and, according to its present owner, Lorenzo H. Zambrano, of Monterrey, Mexico, was intended to be used personally by Nuccio Bertone. Zambrano says, 'after the car was shown twice, Bertone sold it, much to Sig. Ferrari's dissatisfaction as he had consented to sell the chassis to the coachbuilder for exhibition purposes –

*Steve Tillack's impeccable restoration of 3269 GT's Lusso engine compartment evidences the sort of attention to detail that characterizes the finest work money can buy. Details include conventional air cleaner, twin oil filters on this later SWB motor*

on the condition that the car would not be sold privately.' Nice story. But this is not now thought to be true.

Originally, the car was dark blue, but Bertone later changed this to silver. Recently rebuilt by Steve Tillack, who restored both Bertone SWBs, the car was shown at Pebble Beach in 1982 and attracted a lot of admirers. Many people consider it the most striking of the special-bodied SWBs.

Before leaving the special *carrozzerie* SWBs, it's essential to mention the second 250GTO prototype. Jess Pourret now confirms this to be *2053 GT* and not *3175 GT* which was used for engine testing. This car, looking like a plaster-of-Paris hybrid of SWB and later GTO features, was tested extensively at Monza as Ferrari's engineers and technicians strove to perfect the aerodynamics that would allow his next series of GT cars to easily exceed the 160 mph barrier. Known as 'the Anteater' for its long nose, the car's rather crude aluminium body was built at the Ferrari factory. Stirling Moss was one of the last testers of this car in September, 1961. Remember, Moss had logged many successful hours behind the wheels of two of Rob Walker's SWBs, so he was an ideal person to see if the latest Maranello

Left and above 1737 GT; *an SWB that's not an SWB? It's the ex-Peter Sachs short-wheelbase spyder, which was built before the first SWB California. On a specially shortened and lightened 250 chassis, Pininfarina built this styling exercise to see how the cabriolet would perform. The engine is built to full competition spec. The car has often been erroneously identified as either an SWB or a 400 SA Spyder. Photos Peter Sachs*

*Probably the best known rebodied SWB is the famous 'breadvan' or camionette. Count Volpi di Misurata of Scuderia Serenissima commissioned Giotto Bizzarrini to build this car when GTOs were in short supply. The engine was repositioned back in the chassis and a 6-carburettor dry sump engine was fitted. As the GTO's five-speed gearbox was unavailable, the team had to make do with a four-speed SWB transmission. Here is Carlo Abate at Brands Hatch in 1962. Photo E. Selwyn-Smith, Tim Parker Collection*

experiments had improved the breed. Sadly, his accident, the following spring, precluded any chance of his competing in the new 250GTO. The *2053* was raced in standard SWB form after the experiments. Sadly, it was destroyed in a crash at Spa in 1964. *3175 GT* was rebuilt as an SWB after its test period and competed at Daytona as late as 1964 where it finished 12th overall.

A few SWBs were rebodied by contemporary Italian coachbuilders. However, two cars often thought to be SWBs, *1623 GT* and *1777 GT*, were really built on shortened 250 PF coupé chassis by Neri e Bonaccini in Modena to the specifications of Tom Meade, an American living in Italy, who became known for his exotic designs. Both of these cars are attractive – *1777 GT* resembles the later 275GTB with lengthened nose, enclosed headlights and an abbreviated tail. *1623 GT* was a 'chopped' 250 with a front end resembling the 1964 250GTO.

Another short-wheelbase 250 that's not an SWB is *1737 GT*, called a Cabriolet Speciale by former owner, Ferrari enthusiast, Peter Sachs, of Stamford, CT. Sachs advises, 'This is really the first 250 short-wheelbase open spyder, having been built before the first SWB California.

'The car incorporated a specially lightened chassis with stiffening braces. The all-alloy body was built by Pininfarina in plain 400SK style for the original owner who wanted something light and fast. The car was built to full competition specs with a 280 bhp motor, Testa Rossa camshafts and a racing gearbox.'

Sachs indicated this spyder never had a soft top, but was fitted from the beginning with a fibreglass hardtop incorporating a moonroof. First exhibited on the PF stand at the 1960 Geneva Show as an open spyder, without bumpers, it was later shown with the hardtop in place and bumpers installed. It's pictured in the

*Chris Kerrison's ex-Moss SWB rebodied by Drogo. Actually better looking than this photograph suggests. Photo E. Selwyn-Smith, Tim Parker Collection*

*While some SWBs were bodied to resemble Superamericas, at least one 400 Superamerica received an SWB-type body. Here's 3673 SA, virtually indistinguishable from a late SWB. Photo Kurt Miska*

Fitzgerald and Merritt book where, Sachs advised, 'it's incorrectly identified as a special-bodied 400 SA.' Purchased new in France, the car went through a succession of French owners before Sachs bought it in 1979. He restored the car to full original specification and sold it to Charles Johnson, Jr., Miami, Fla.

Neri e Bonaccini also rebodied a Lusso SWB, *3771* GT, as a 'Nembo Spyder.' Again, the prolific Tom Meade styled the rebody – a car whose SWB origins make it an interesting collector's item today. The car is owned by Fred Jaeger, Racine, WI.

In 1967, Piero Drogo rebodied *2209 GT*, a competition SWB, with a flowing body design by Tardini reminscent of the Jaguar E type. Jess Pourret reports that this car's angular lines and enlarged glass area are its distinctive features. As well, the competition engine was replaced by *4921 GT*. A similar Drogo/Tardini rebody, on Lusso SWB chassis *2067*, has Cromodora alloy wheels from a 365GT. A sister car, also in France, *2067 GT* is a Tadini rebodied Lusso equipped with Borrani wires.

Probably the best known SWB rebody effort was that of ex-Ferrari engineer Giotto Bizzarini on an SWB competition chassis *2819 GT*. This was the famous 'breadvan' or *'camionette'* built for Count Volpi di Misurata of Scuderia Serenissima. After the famous split that saw engineers Chiti, Bizzarini, Tavoni and Gardini leave Ferrari in 1961, Bizzarini's talents were available. Volpi already had 250GTO *3445 GT*, but he wanted a second car. When GTOs were seen to be in short supply, the enterprising Count simply had Bizzarini modify a Comp./61 SWB to suit.

*2819 GT* was an ex-Olivier Gendebien car that had finished second in the 1961 Tour de France. Under the ex-Ferrari engineer's guidance, the car received a dry sump modification, six Weber 38 DCNs and the engine was moved backward and dropped lower in the modified chassis. The limited-production 250GTO five-speed gearbox was unavailable, so the team had to make do with the SWB's four-speed. Even so it was a very quick car.

Bizzarini designed a new body, using the latest adaptation of Kamm aerodynamic theory. The car built by Neri e Bonaccini, ran at Le Mans in 1962 where, due to some over-enthusiastic driving, it broke and DNFed. The chopped-off body, with its long nose, was called *'camionette'* by the French, but it's more popularly known among *Ferraristi* as the 'breadvan'.

Later, the 'breadvan' was fitted, by Count Volpi, with rear bumpers and a rear window wiper for road use. It's presently in the US and was exhibited by its owner, then Gordon Tatum, at the Ferrari Club of America annual meet in August, 1984.

For 1963, yet another Drogo rebody appeared at Le Mans, built on competition SWB chassis, *2053 GT*. This car was originally used by the factory for GTO experiments and may have been the 'Anteater'. Pourret's research indicates that after the first rebodied SWB appeared at Le Mans, the concept of rebuilding a crashed SWB became quite popular – especially as 250GTOs were in short supply and this was certainly a less expensive alternative. Drogo, with help from Neri e Bonaccini, who certainly knew their way around the

*Another Drogo rebodied SWB for Scuderia National Belge shot in 1963. Photo Unitas, Tim Parker Collection*

SWB by now, began on *2053 GT.*

This car was modified with GTO heads, six Weber 38 DCNs and a repositioned engine. Pourret commented '...the shape was pleasant enough, but the front end was marred by two ugly openings for the radiator air intake.' Sadly, this SWB suffered a series of misfortunes. Damaged while returning from the Angola GP, it was rebuilt, only to be raced unsuccessfully in 1963, and, denoted earlier, it was subsequently destroyed at Spa, the following year.

*2735 GT* was a 290 bhp Comp./61 SWB owned by Rob Walker and raced for him by the legendary Stirling Moss. The car was crashed by a subsequent owner, Chris Kerrison, at Goodwood. It wound up with Drogo, and subsequently, Neri e Bonaccini as Kerrison wanted a modified rebuild. The car was lowered, lightened and fitted with a dry sump and the now ubiquitous six Weber 38 DCN manifold. Kerrison campaigned the car in 1963–64 until it was wrecked in Ireland. Rebodied today, and fitted with a modified engine from *3635 GT*, its owner lives in England. Interestingly, the original modified engine used to live in 250GTO *3729 GT.*

A third car, *2439 GT*, was sent to Drogo by its owner, Swedish driver, Gunnar Andersson after a bad wreck in 1962. Surprisingly, Andersson changed his mind about the rebody and the car was subsequently shipped to Scaglietti where it was fitted with an original SWB-type body.

While many people were changing Superamericas to resemble SWBs – and SWBs to look like GTOs, at least

*2231 GT was fitted, at some point after initial construction, with an attractive twin-nostril Carlo Chiti front end reminiscent of the then-contemporary Ferrari Formula 1 cars. This car was originally owned by Adrian Conan Doyle; it's now owned by UK Ferrari enthusiast, Bill Lake*

one 400SA received an SWB lookalike body. This car is *3673 SA* and its main distinguishing feature is that the gas filler cap is located externally on the right-hand rear fender.

Finally, a number of SWBs (see appendix) were modified slightly. *2111 GT* was originally a 1960 car rebodied as the prototype for the 1961–62 SWBs. *2231 GT* was fitted with a custom – and quite attractive – twin nostril Carlo Chiti-type nose. *2639 GT* had faired-in headlights added, in the early 1970s after a California highway accident. *2649 GT* reportedly also has covered lights. *2839 GT* once had body modifications resembling a 275GTB and *3359 GT* incorporates taillights similar to the 250GT Lusso/275GTB. It should be noted that *2501 GT* and *2521 GT* were both rebuilt after highway accidents to 'close to SWB' specifications.

The Ferrari saga is replete with interesting trails and dead ends, as historians and enthusiasts attempt to piece together the tangled trail of variants. I'm sure, while carefully researched, that this listing may be incomplete, but perhaps, with the publishing of the book, we'll hear from others who have special-bodied SWBs. As such, Ferrari's intriguing history continues to become even richer.

# Racing highlights

SWBs distinguished themselves in a wide variety of major races – and many, many club events – from the outset.

The TDF 250GT was a true competitor in its day, but subtle improvements in the SWB, emphasizing more power and better handling, gave Ferrari a continued advantage. The proven V12, the addition of powerful disc brakes and the benefit of the shortened wheelbase all combined to wreak havoc on the competition.

In March, 1960, Ed Hugus and Augie Pabst began the SWB's distinguished career with a fourth place, in *1785 GT*, at the 12 Hours of Sebring. The very first SWB, *1539 GT*, driven by Sturgis and Grey, took a sixth for Chinetti's North American Racing Team.

Two months later at the Targa Florio in Sicily, *1791 GT*, co-driven by Lenza and Maglione, finished ninth. Sorting out exercises for these new cars were very much in evidence as Beurlys-Noblet drove *1811 GT* to a 15th place finish at the Nurburgring on 22 May, 1960.

The SWB's record quickly began to improve. Early in June, Seidel and Dumay finished the eighth GP of Rouen in third and fourth place, respectively, Seidel drove *1807 GT* and Dumay's car, number is not known.

Contenders now, SWBs finished the Le Mans classic in fourth to seventh places – with Tavano/Dumay in *2001 GT* leading the pack. The *1759 GT*, found its way into seventh place with Ed Hugus and Augie Pabst once again sharing the driving.

Thoroughly warmed up, *2119 GT* distinguished itself with the SWB's first major win in the 25th Tourist Trophy at Goodwood on 20 August. Barely a week later, the victory was repeated with a Redex Trophy win at Brands Hatch. Both times, the blue Rob Walker Equipe car was driven by the legendary Stirling Moss – a man who was very comfortable in SWBs.

Moss talked about his SWB experience in a chapter of Stan Grayson's excellent book, *Ferrari, The Man, The Machines*, saying '. . . the Ferrari I came to drive most often was a hotted-up 250 GT Berlinetta with a 3-litre V12 pushing out more than 300 bhp (sic). In its first season, this car helped establish Ferrari domination in races like the Targa Florio and Le Mans and brought home the World Championship . . . And it certainly did me proud. Three-litre Ferraris took me to wins in the Tourist Trophy races at Goodwood and Nassau in 1960

*The master, himself, Stirling Moss, is shown at speed in Comp./61 2735 GT. Moss drove two SWBs for Rob Walker and won convincingly at the Goodwood TT and Nassau, in the Bahamas. Drifting the berlinetta, concentrating intently, Moss' great skill would have been well utilized in the later GTO. The popular British driver tested the GTO prototype, but, sadly, after his grand prix accident driving a Lotus, he was unable to compete again. Photo Geoff Goddard, courtesy of Lowell Paddock, and* Automobile Quarterly

and again in 1961, and victories at Brands Hatch and Silverstone.

'In my final racing season, 1962, we might have done better still: in the three-hour Daytona Continental, I won the Touring Class, coming in fourth overall, two laps behind the leader. And in the Sebring 12 Hour race, Innes Ireland and I were doing quite well, with fastest lap to our record, when an early refuelling stop led to our being disqualified. With a record like that, over five seasons with interruptions, you can't help thinking highly of a car.'

Moss wasn't alone, of course. For other drivers in SWBs, the wins kept coming. In September, 1961, Carlo Abate drove *1999 GT* to first place in the Coppa Inter-Europe at the very demanding Monza track.

But, the best was yet to come. In its first Tour de France, September, 1960, the SWB proved its mettle in the type of driving it was best suited for – Willy Mairesse

*Rob Walker's second SWB for Moss, this time without the radio. Legend has it that he listened to the commentary on his Goodwood TT win as he drove along! Photo Patrick Benjafield, Tim Parker Collection*

and his co-driver, Berger, upheld the TDF's honours for Ferrari in *2129 GT* with a first overall. They were followed in second and third by Schlesser/Dumay (*2127 GT*) and Tavano/Martin (*2001 GT*). Olivier Gendebien and Lucien Bianchi, thought to be real contenders, pushed too hard and blew a piston. Hurriedly rebuilt by the Florini SEFAC team, Gendebien's car (*2149 GT*) limped into 14th place in the French classic.

Bianchi, teamed with Gendebien again in a newly rebuilt *2149 GT* were back in fighting trim a month later leading Schlesser/Simon (*2209 GT*) and Dumay/Tavano (*2127 GT*) in a 1-2-3 sweep of the 1000 km race run on the famed Montlhéry circuit outside Paris.

December meant Nassau, the racing party in the Bahamas that combined warm weather frolics with serious track work. As mentioned, Stirling Moss was there with Rob Walker's *2119 GT* to take the Nassau Trophy convincingly. The following March, Denise

*Denise McCluggage watches Allen Eager at Sebring in March 1961. One of the best of the early US women sports car drivers, McCluggage and her co-driver, Eager, finished tenth overall, first in GT class in 1931 GT. The SWB finished fifth OA, fifth in GT at Le Mans the previous year co-driven by Arents and Connell. Photo Denise McCluggage*

Above *Getting ready for the Le Mans start at Sebring, co-driver Allen Eager helps Denise McCluggage with her helmet. The pit scene looks quite frantic with everyone making last minute preparations before the start. Photo Denise McCluggage*

Right *Denise McCluggage in 1931 GT at the Nurburgring in May, 1961. McCluggage had driven the car from Italy to Germany, then raced successfully in the 1000 Km event until co-driver Eager stuffed it in the Karussel turn*

McCluggage and her co-driver Eager took a tenth overall, first in GT class win at the Sebring 12-Hour endurance GP.

I spoke recently with Denise McCluggage who remembered well her SWB experiences. 'I drove at Sebring with Allen Eager in 1961', she recalled. 'We finished tenth overall – and won the Gran Turismo category. The car, *1931 GT*, had previously belonged to George Arents and he finished fifth overall in it with Alan Connell at Le Mans in 1960. This berlinetta wasn't just a race car, you understand, it was a road car, too, painted powder blue – one of the last that could do both well. The three carburettors made it wonderfully tractable.

'The SWB was a superior machine with superb handling and wonderful manners. I had to sell it. You see, it was not only my only *car*, it was my only *thing*. Crashing it would have wiped me out.

'I became aware that the car had to go at a race at Thompson (CT). It was a preliminary ten-lapper and I was wheel-to-wheel with Dick Thompson driving a Corvette. We were headed into the first turn and I knew he'd chop me off if he could. I thought: "I can't bend this

*Bob Grossman was an American racer who had a lot of success in SWBs. Here he is at Marlboro, Maryland, in 1962, with the chequered flag after an A Production victory. The car is Comp./61 2731 GT; Grossman had driven it to a sixth OA, 2nd in GT, at Le Mans the previous year.*
*Photo Alan Boe Collection*

Above *3005 GT was a very quick Comp./61 alloy car with an interesting history. Originally owned by Luigi Chinetti and his North American Racing team, this car, driven by Pedro and Ricardo Rodriguez, won the GT class at the 1000 Km of Montlhéry in 1961 – and took the Index Award and, as well, turned the fastest lap in the same event. Later sold to D. Thiem, it enjoyed a successful SCCA career. Here, it's shown at Marlboro, Maryland, where owner Theim won a third OA on his way to winning the 1962 SCCA class championship. Photo Alan Boe Collection*

Right *At the Nurburgring in 1961, the Scuderia Serenissima 2733 GT is shown on its transporter. A*

thing!" I braked hard but Dickie still hit the left front headlight rim and sent it spinning. Clearly, I could not race my apartment, my winter coat, my all. After the race I went over to Bob Grossman and said: "You've been coveting my car. Well, it's yours." And I went home. Didn't even race in the finals.

'Earlier that summer, I had a terrific race with another Corvette. This time it was at Meadowdale (Ill.) with Don Yenko. Yenko flew off and left the rest of the 'Vettes. I chased him, setting a new lap record in the process.

'We had a helluva dice, but Don always drove mostly sideways and I couldn't get by him. He churned up some sand from the side of the course and I lost the Ferrari in a big way – just where the course went through a sort of ravine. I didn't want the car to climb the sides and roll so I spun it – four or five times – down the middle of the road. The engine died. It must have been 30 seconds before I got it started again. Still no other cars caught up. That's how far ahead we were. The upshot was the other Corvette drivers protested Yenko. He was found slightly over-inches so I ended up official winner.

*Comp./61 lightweight, driven by Carlo Abate, the car finished first in the GT class on this occasion and turned the fastest lap. Reportedly, this car was badly wrecked in the 1970s and is currently equipped with engine number 1733. Photo Karl Dedolph*

*A nice period shot of an SWB Comp./61 lightweight, it's shown in the parking lot at Bridgehampton, LI, in August, 1961. With New York license plates fitted, the car, believed to be 2731 GT before it was repainted red, could easily have been driven to the track, raced and then driven home. Note all the big American iron in the background – the SWB was a street sensation in the early 1960s. Photo Kurt Miska*

Above *Photographed at the Monterey Historic Races at Laguna Seca, this early SWB sports Snap exhausts, Italian racing stripes and a very special license plate. Interestingly, SWBs are still competitive in vintage racing and many owners are competing. A close look shows this car has been fitted with a non-standard rollover bar. Photo Marc Tauber*

Right 3327 GT *fitted with one inside and one outside right side gas filler, is shown at Cumberland, MD., in 1964*

'The SWB was marvelously reliable. One of the best Ferraris ever. Even Bob Grossman said it was one of the best Ferraris he ever owned and he had owned a passel of them.

'Allen Eager and I had a pet name for the car. Allen was a jazz musician and musicians call cars "shorts", who knows why. We called the berlinetta our "short short".

'With the Sebring money, we flew the car to Europe. We went to Taruffi's driving school in Modena then drove to the Nurburgring for the 1000 Ks. Spent a week on the course practising before the race. Great fun. Unhappily, Allen popped out of the Karussel during the race and hung the car on a fence. Afterward, we pried the fenders off the tires and drove back to Modena. I tell you nothing could get you more attention in those days than driving a damaged Ferrari with the shadow of race numbers on it across the Continent.

'We took the car to Scaglietti for some of that great Italian eye-ball bodywork. Beautiful. And repainting. I had it painted an electric blue with a wide white racing stripe. That car was so blue it was red!

'After the Thompson affair I turned the SWB over to Bob Grossman for $6000 and a used Mini. (I had paid

2937 GT, *now owned by Pierre Bardinon flashes past in a vintage race. Co-driven by Mairesse and Berger, this SWB won the 1961 Tour de France. Equipped with oversized Weber 46s, 16 in. wheels and special competition exhaust, the Comp./61 lightweight had a distinguished racing record. Photo Pierre Bardinon*

Le Mans 1961: at least four SWBs chase right from the start. Also in the photograph are a works Testa Rossa, two Aston Martin DB4GT Zagatos, a couple of Maseratis and more. Photo Tim Parker Collection

Luigi Chinetti $9000 for the Ferrari before Sebring.) Seemed like a fair enough deal at the moment. But last time my old SWB changed hands I heard it went for over $100,000. Well, I was born fast, not quick.'

Overseas, as well, SWBs continued their winning ways. Mike Parkes and Peter Whitehead, in *2119 GT* and *2009 GT*, were first and second at Snetterton in March – and Parkes triumphed at Goodwood the following month. Successfully campaigned in many races, major and minor, the SWB's feats would fill an entire record book. Many of the key events are highlighted in the appendix – and a few more of the most important wins should be mentioned here.

# What's it like to drive?

In the late 1950s, Ferrari road tests were usually done with privately owned cars. The distributors weren't too inclined to loan out their precious stock to a group of enthusiastic road testers – particularly when Ferrari was only making a small number of street cars each year – and those buff book readers who could afford one of these hyper-expensive toys represented a small minority.

As a result, road test subjects were infrequently available and many of the top Ferrari performers of the early days were never really 'tested'. The British magazines were sometimes an exception – after all, road testing had originated in the UK, and some of Britain's motoring journalists were considered very trustworthy, competent drivers by the Ferrari factory as well by the local distributors.

Maranello Concessionaires' owner, Col. Ronnie Hoare, was the right sort of bloke, so he happily let *Autosport*'s John Bolster 'borrow' a trio of Ferraris in September, 1960 – a Pininfarina 250GT coupé, a 250GT SWB and a V12 Testa Rossa. Turned loose at Goodwood, Bolster's enthusiastic impressions of the light blue SWB (actually *1993 GT*, the first steel-bodied and the first right-hand drive car) are worth recalling:

'This is an outstandingly easy car to drive,' Bolster reported, 'and the short wheelbase renders it noticeably handier on corners. Even very fast cornering is completed without drama, and one can remain totally relaxed while really sliding through the bends. Once again, I tried the machine on road speed and racing

*Autosport's John Bolster tested an SWB in September, 1960, calling it '. . . an outstandingly easy car to drive . . . and the short wheelbase renders it noticeably handier on corners.' Shown is 3477 GT, rebuilt from a badly damaged basket case by owner, Herr Fehlman, in 1974–75. Photo Marcel Massini*

tyres, and though the racing tyres were noticeably better, the difference in handling was not so marked as with the Pininfarina (250 GT Coupé).

'The berlinetta is a superb competition car,' the tester continued, 'but it is almost ideal for fast touring. Its small size is a great advantage, and the equipment and creature comforts are in the luxury class. . . . The 12-cylinder engine, even when really 'hot', is smooth enough to render this one of the fastest and most desirable of road cars.'

Similarly, the editors of America's *Sports Cars Illustrated* (now *Car and Driver*), drove an SWB, in September, 1960, fresh from the races. The car was Charlie Kreisler's *1785 GT*, the fourth place finisher at Sebring that year, driven by Ed Hugus and Augie Pabst.

The oft-quoted road test was certainly a labour of love.

While I'd like to reprint it in its entirety, space considerations require some judicious editing. Nevertheless, the keen enthusiasm from 25 years ago still rings true:

'From the tips of its thrusting headlights to the end of its tucked-in tail, Ferrari's newest berlinetta exudes the essence of speed and power. For once the externals don't mislead. This is a fast car, potent almost to the point of being brutal . . . if this Ferrari seems bigger-than-life at rest, it's definitely so when on the go.'

The *SCI* staffers, used to a wide variety of fast cars, judged the Ferrari in the top ranks. 'Right now, we want to establish, without equivocation,' they stated, 'that it's powered by the greatest automotive engine in the world today . . . (an engine) that has lived for 13 years, almost constantly in production in different sized variants, and always able to win major races. . . .'

With that for starters, the admiration had just begun. Behind the wheel, *SCI*'s testers enjoyed the SWB's quickness: 'Throttle response is live and obedient to a toe's touch,' they wrote, 'above the berlinetta's 1400 rpm green-line. If you stab it suddenly below that speed, either in or out of gear, there'll be a pause of throat-clearing before the action starts. When you take your foot off the pedal, there's a quick response too, a rapid drop in revolutions according to your needs. You'll get an idea why when you switch off, too: The engine halts with a sudden "whump" that betrays the absence of excess rotating mass.'

Underway, the editors' enthusiasm increased as the revs rose. 'When it's running, this ultimate engine is exhilarating, electrifying. Twelve pistons, 24 valves and roller-tipped rockers, two cams, a few yards of chain and an assortment of pumps combine to produce the most wonderful racket ever to reverberate in an eager enthusiast's eardrums. This soul-stirring cacophony is controllable – the first-time driver learns with a thrill – up to and including a gut-quivering roar with only gentle pressure of the right toe.'

From the driver's seat, the thrills didn't diminish, either. 'In the cockpit, the din is muted to a many-levelled mechanical whine up front and an urgent crackle from the quadruple tailpipes . . . innocent of insulation, the Berlinetta's hood and firewall are built for lightness, not silence.'

One of my favourite comments ever about the SWB came next in this road test, from an unidentified lady passenger. 'It's instant car,' she said. Not surprisingly, the editors, accustomed to all manner of fast machinery, had met their match. '. . . we must admit that this 250GT Ferrari has more sheer acceleration than you can use on the highway for more than a few seconds at a time. At whatever speed you're travelling, those few seconds, at full throttle in the proper gear, project you so violently into another speed spectrum that you have to reorient yourself completely.'

At the other end of the rev range, the *SCI* editors marvelled at the SWB's excellent manners around town. 'Pottering along at 40–45 mph in top gear, the berlinetta gurgles and rumbles just below the 'step' in the carburation, erratically but not annoyingly. There's no

*'Ferrari's newest Berlinetta exudes the essence of speed and power,' said* Sports Car Illustrated *(now* Car and Driver*) in 1960. 'For once the eternals don't mislead. This is a fast car, potent almost to the point of being brutal.' Shown is 3175 GT, owned by Giuseppi Lucchini. This car was used for GTO studies by the factory, and equipped with 46 mm Webers. It was later rebuilt as a street SWB. Photo Marcel Massini*

danger of plug wetting, and just a tap of the treadle is needed to arouse the dozing cylinders.'

With road speed limitations, even in the pre-55 mph days, a handicap for the fast Ferrari, the magazine's staff exercised the racer at Lime Rock Park for some real thrills. One of the passengers reacted like this: 'You glance at the speedometer; its needle hovers over the 110 mph mark down the straightaway. There's wind noise, engine noise and exhaust noise. The timing stand blurs past. You feel an uncomfortable lurch as you brace against the heavy deceleration for the first corner. The throttle is cracked and you are pressed back and to the side unmercifully by brutal torque. A rush of hot, almost stifling air blasts back from the engine and smells of gasoline and hot rubber form a heady mixture. Through the corners, you slam from side to side in the seat; the tach needle dances near its upper limit then plummets with the upshifts. Your driver sweats from effort and engine heat as he works over the wheel and gear shift. Before you realize it – in just the short time you've been watching what's happening in the cockpit – you're "lost" on the course and it takes a moment to get your bearings again. Things happen fast in a Ferrari! You sweep into the downhill turn, tires scrubbing for the last ounce of adhesion, the rear end bobbles a little, and with a roar you streak down the straight again.'

Enthusiastic? Of course, the test concluded with a summation that very capably describes why the SWB remains the classic dual-purpose racer. 'We ask this question,' the testers challenged, 'what other automobile anywhere in the world, for any price, can do what this car can? Can you conceive of another vehicle that, on one hand, is a virtual cinch to finish in the top five in a major international sports car race? Even with an unlimited budget, do you think you could come up with a design that would blend these conflicting requirements as successfully as the 250 GT berlinetta does. Give it some thought. We have, and we've come to the conclusion that this car is not only a bargain at $14,000;

it's an absolutely unique piece of machinery . . . the finest genuine sports car we have ever driven.'

*Autocar* tested an SWB in May, 1964. The editor borrowed a left-hand drive car, *1917 GT*, driven to a GT win at the Nurburgring in 1962 by the team of Nocker and Seidel. Geared for hillclimbing the car was limited to a 130 mph top end, but with blazing acceleration. The solid axle certainly wasn't considered a limitation here. '. . . the surprising and satisfying thing is the way the car surges forward without a trace of wheel hop and with a great reluctance to spin on the part of its big tires. If this can be achieved with a really good live axle,' the tester pondered, 'why get involved with independent suspension and its possible vagaries?'

On the road, distances were spanned in record time. 'A car with this sort of performance covers journeys in England very fast, mainly because it is seldom delayed. The smallest gap gives ample time to overtake slow or parked vehicles. In addition, of course, corners and bends seem to be easier to round and short straights allow bursts of 100 mph.'

This was only the editor's second time in a Ferrari, and he admitted the time was too short to enable him to handle it to its capacity. Nevertheless, showing what a tractable car an SWB is, the driver noted, '. . . the car feels short and neat, and occasionally I felt brave enough to accelerate around a bend in second, at the same time paying off the steering in anticipation of the slide that would otherwise be induced . . . it was almost as if the Ferrari looked back at me and said, "that's better, now you are getting the idea." That car really wants to go and it cooperates with a driver who has the same idea.'

The SWB's four-wheel discs were impressive. *Autocar*'s editor noted: 'There is no fun going if you cannot stop, but I was agreeably surprised by the brakes . . . they are powerful and consistent, and the pressure required is never particularly high.' On the downside, the clutch came in for some criticism. 'For a competition car, the clutch is also quite light and progressive. It hated the least bit of slip and quickly protested by filling the interior with a horrid smell of hot lining.'

*Cars & Car Conversions* borrowed the same Ferrari SWB in December, 1965. By now, most of these cars, obsoleted by the GTO, were enjoying second careers as club racers and fast street machines. The SWB's new owner, George Burton, fitted the car with a few amenities for road use – a Lucas alternator, iodine vapour spotlights, and glass side and rear windows.

The editors were very respectful of a borrowed car, keeping the revs to 7500 rpm and commenting wistfully that '. . . the only thing we can afford (was) . . . the spotlight covers.' Some of their comments on the SWB's interior were equally amusing. 'There is even a clock and unlike the Rolls-Royce, you definitely can't hear it ticking at 60 mph. . . . Really, the interior is surprisingly well upholstered and trimmed for a competition model, with a conspicuous lack of the mock Chippendale which mars the interior of so many UK cars.'

Underway, the testers rediscovered what the SWB was all about. 'With 280-bhp available . . . the road performance is shattering, not only in the top gear which

*'You glance at the speedometer; its needle hovers over the 110 mph mark down the straightaway. There's wind noise, engine noise and exhaust noise. The timing stand blurs past.' Another enthusiastic quote from* Sports Cars Illustrated: *the car is 2765 GT, owned by Edoardo Genuizzi. Photographed at Imola in 1983 by Marcel Massini*

*1917 GT, shown in the foreground, was the* Autocar *road test car in May, 1964. A former Nurburgring GT winner, the SWB was still capable of thrilling road achievements. 'A car with this sort of performance covers journeys in England very fast,' the testers wrote in an understated way, 'mainly because it is so seldom delayed.' The SWB in the background is 2731 GT. Photographed at Dijon, France, by Marcel Massini*

is virtually in the "low-flying" category, but in terms of acceleration, handling and braking, which make very high average speeds possible, even on British roads. It would be unwise to admit the figures achieved in case the Minister of Transport hears of them . . . the car was astonishingly agile round the country lanes, taking the sharp bends as though on rails. . . .

'Fast cornering is a matter of setting the car up on a chosen line and holding it there with the throttle until the apex is reached, when the power can be turned on to bring the tail round, simultaneously chopping on a twitch of corrective lock to point yourself up the next straight. Needless to say, there is no body roll, the car remains rock steady. It also remains rock hard; the dampers are set for competition work and if you go over a penny, you can practically tell if it's heads or tails.'

Like roadtesters before them, the editors excitedly described the SWB's performance at speed. '. . . as the engine approaches peak revs in the gears, the noise rises to that distinctive Ferrari howl, drops for an instant as you change up and then comes in again with a tremendous bark as you accelerate in the next gear. A real "Mulsanne straight" scalp-tingling noise if there ever was one.

'Nothing is perfect,' the editors found, however. 'There are one or two disadvantages, like no luggage space – the boot being occupied by a large tank which worried the garage man when he filled it; "I've put 24 gallons in, guv – could it be leaking somewhere?"'

Topped up and at full chat again, the SWB proved it was '. . . one of the safest cars on the road, provided that the performance is used with discretion. There is a tendency to find the roads littered with small cars hurtling towards you backwards, but apart from certain gentlemen using the fast lanes of motorways to set their own speed limit, most drivers pulled over very swiftly and courteously to let us through. But then, who wouldn't?'

Twenty years later, three contemporary accounts of SWBs revisited appear to underscore the earlier findings.

A few years ago, *Thoroughbred & Classic Car*'s editors drove an ex-Le Mans SWB, owned by Simon Philips, from the UK to the Sarthe circuit for the race. The SWB, *2009 GT*, was the ex-Graham Whitehead lightweight car with an interesting if not distinguished competition history. Rebuilt, but not over-restored, the Berlinetta was an exciting way to get to the races.

'What's driving such a car in 1981 like? The answer is, "surprisingly civilized." The Berlinetta is an odd mixture of the normal in road cars and the spartan in racing cars . . . it's when you start up that you realize what sort of car this really is. It's not embarrassingly noisy by competition-car-on-the-road standards, but nor is it quiet . . . the exhaust can rise to a respectable bellow, with intake noises in sympathy – but it is the purely mechanical noises which are remarkable. Whether cold or hot, the clatter is disquieting, to an extent that makes the car sound like a very much older machine designed

*Somewhere in Europe, an unidentified SWB dashes up a road – exhausts trailing a characteristic whiff of blue smoke. The music from those four Snap exhausts enchants all who listen . . . if you stare at the photo intently, you can hear it too. Photo Karl Dedolph*

at least three decades earlier than 1960. It is as if there is a big complicated mechanism in front of you, containing lots and lots of little tapping parts.' Certainly, that's not an inaccurate description of a Ferrari V12.

'The engine will snuffle away at low-ish speeds, when the clashing knitting needles are loudest, and the car is being trickled through town or village at around 1500 rpm. The noise is still like driving behind a large bag of nails at up to 3000 rpm; exhaust and intake take over from there onwards.

'Ride by today's standards is poor, but by no means bad compared with other cars of this type, and it does get better as you go faster. . . . It steers well by older standards, which is to say it is tolerably accurate, good in straight stability, on the heavy side and prone to falling down cambers, with correspondingly good feel. The big Dunlop discs work well, and the machine's tool-like gearchange is a delight – not light to use, but with good precision.'

Artfully, the travellers liked the SWB's restrained appearance: '. . . it has a bound-up crouch to its side view, which to some eyes makes it look like a car seen through the vertically-travelling focal plane shutters of early motor racing photographers, giving old racing cars a leaning forward look.' Travelling through France, greeted by waves and flashes from the other motorists made the SWB time travellers smile. 'Frenchmen, Italians and Germans are vastly more car-minded and pro-car than the average Briton,' the testers mused, 'who in the mass accepts, uses and cleans his car, but does not love it.'

*Thoroughbred & Classic Car*'s Tony Dron drove a freshly restored SWB at Goodwood for the January, 1985 issue. Owner Roy Pearse had just paid for a comprehensive ground up restoration by Ron Beasley and while discretion was necessary with the newly rebuilt engine, Dron was still impressed. 'The racing version of the SWB is, in my personal opinion, the second most beautiful car produced yet, the most beautiful being the Ferrari P4.'

Dron, a Porsche enthusiast, candidly confessed, '. . . when it comes to stirring the emotions, Ferraris get my heart churning.'

*2163 GT*, now fitted with engine *1615*, not an SWB motor, underwent a very detailed restoration, with a number of its alloy body panels requiring re-fabrication. The car was raced originally by Scuderia Serenissima and after a DNF at Montlhéry in 1960, it bounced back to win the 1000 km at the Nurburgring the following year, driven by Abate and Davis.

Rolling up the radiator blind to raise the water and oil temperatures, Dron cautiously circuited the track, carefully increasing speeds, '. . . a few laps later and we raised the rev limit to 4000 rpm which had the car moving along quite fast by any standards, though it still wasn't really necessary to lift off for St Mary's corner. Already, one could enjoy some sensation of speed, the delicious noise from the Columbo V12, even on such small throttle openings, and the marvelous feel from the reasonably light steering. The large woodrim steering wheel and perfectly placed big aluminum gearknob, which is so high it seems almost level with one's shoulder, are a real pleasure to use. The controls may look big and cumbersome, but in use they are so smooth that they flatter the driver and immediately give that deep sense of satisfaction that one derives from operating superbly engineered machinery. . . . I was left with the feeling that I wanted to stay with this car, see it through its running-in period and have a real go to see what we could get it to do in full cry.'

Lastly, in *Road & Track*, former Grand Prix champion Phil Hill recently drove a competition SWB in retrospect and on the same occasion, drove a modern 308 Quattro-valvole. The SWB was *2083 GT*, retrofitted with wind-up windows and vent wings, and fitted with a six-carburettor manifold.

Hill enjoyed the older car, a model he'd never tried before. 'It feels like a Ferrari,' he said, 'and a good one, too. It runs beautifully. When you do things wrong, you can fix it easily. It starts to go into a long understeer. If you know what it's going to do, right when it begins you can lift off and it gets pointed correctly.'

Hill had a wonderful time trying to provoke the SWB into a spin. 'I tried everything,' he said, 'but I just couldn't do it. You've got to yank your foot off to get it to take a new point.'

The verdict – in a comparison of old racer and new street car, Phil Hill commented, 'The good road car of today is as good as the race cars of yesterday.'

Perhaps, as the definitive statement for the SWB, we should examine the converse of that remark. Today, the modern Ferrari, while a formidable street performer, needs to be extensively modified for competition. In the SWB's era, though, a high performance car had to perform both roles handily – and the Berlinetta did just that. As a dual-purpose GT, the SWB knew no equals – today, it's nice to know the old warrior can still keep up with nearly anything it's likely to encounter on the street. Surely, that's a fitting testimony to the soundness of a great design.

# SWB successors

The SWB was barely a year old in 1960, when Ferrari began making plans for its successor. *2643 GT* was the beginning – a Pininfarina design reminiscent of a shortened 400SA. A radically shaped rear end treatment rounded out this new design. Under the hood, the pattern to come was set with a dry-sump Testa Rossa six-carburettor engine, located well back in the frame. Experiments in body shape and engine modifications were done on other SWBs, *2053 GT*, a competition car, and *3175 GT*, a steel-bodied coupé. These cars also served as a development mule for several 250GTO features. No-one knows for sure which of these cars, if either one, was the famed 'Anteater' test prototype.

For 1962, the FIA established two new titles to replace

*The ultimate development of the 250 GT series was the GTO. Although 100 examples of this car were required under FIA rules, to ensure the winning car was indeed a production model, Enzo Ferrari insisted his new car was a logical extension of the SWB's development. His persistence forced the FIA to homologate the car. Here's GTO 4153 GT, just after winning the 1964 Tour de France, co-driven by Lucien Bianchi and Jojo Berger. After ten years of trying, Bianchi was victorious. Although Andre Simon's SWB beat the GTOs in the Tour de France in 1962, the new cars were to quickly eclipse all previous records. Photo Autopresse International*

the Sports Car Championship. One prize was for proto-types and the second was for GT cars. Qualification for the GT award had an important prerequisite. One hundred examples of the competing car had to be built to 'ensure' it was a genuine production model.

TDFs and SWBs had been consistent GT class winners for several years, so there was certainly the basis for a new effort. However, the seldom complacent Mr Ferrari knew his competitors would be anxious to win the newly defined GT prize. From a marketing point of view, success on the track could be quickly translated to success in the showroom for the winner. Not to be outdone, Ferrari planned for the ultimate development of the 250GT series to ensure the coveted laurels would remain in Maranello.

Actually, as Alan Boe points out, 'over 100 SWBs had been built already and as FIA rules allowed the body style to vary, in fact the GTO had already been homologated. Engine size and other particulars had to be the same, but remember, Ferrari had already catalogued the dry sumps and 6-carburettor manifolds,

*For the street in 1962–64, the GTO was offset by the 250 GT Berlinetta Lusso, an entirely separate, relatively mild mannered GT with lovely styling by Pininfarina. Although a few of these cars were raced by enthusiastic owners, the Lussos were never intended for racing, as the GTO could handily carry the flag under the FIA's revised rules – and race cars began to be much more single purpose machines*

too. Thus, the factory really did not have to build 100 GTOs to achieve official status. They only needed one or two. People often forget this.

Boe also notes, 'there's nothing "official" with Ferrari about the name GTO. It came from the Italian motoring press in an effort to distinguish the new berlinetta from the SWB. Ferrari never tried to "rub it in" with the FIA, as people frequently think – the car under the rules was definitely legal.'

The 250GTO was, of course, the SWB's successor, but only 36 cars would be built with 3-litre engines. Three more were constructed as prototypes with 4-litre engines. Although the two models differed somewhat, characteristically, Enzo Ferrari cleverly built his new car to be a logical extension of the SWB's development. Ferrari also felt that very few privateers had the skills to manage his quick new competition car, so he limited the total quantity built.

The name GTO actually came from Gran Tourismo Omologato – 'homologated', in Italian – and although the factory called the car officially, 250GT Berlinetta Scaglietti Competizione, the shorter name, GTO, became the popular title.

Ferrari had sized up the competition well. They included XKE Jaguars, in lightweight form, the Zagato-bodied Astons and later the quick, big-engined Cobras were also threatening. Porsches, though smaller and in a different class, had good power and excellent handling – and could often outlast bigger and faster cars in endurance races.

Ferrari's strengths lay in the bullet-proof 250 (named for its 250cc capacity) V12 – and experiments with the Comp./61 engines proved additional power could be developed with little effect on reliability. The SWB's tubular chassis, too, with more stiffening here and there, could continue its winning ways. The car's live rear end was a liability, but IRS for berlinettas would continue to elude Ferrari for some time. The Commendatore's emphasis on more than adequate power seemed sufficient.

Aerodynamics were really the key factor for change with the newest Ferrari model. High speeds would be required and the SWB's blunt snout, like the TDF's before it, hit a barn door at speeds above 155 mph (250 km/h). As well, the front end, on an older model, had an alarming tendency to feel light at those velocities.

Ferrari correctly figured that 180 mph (nearly 290 km/h) would be required, so his technicians and engineers experimented for a long while with the aerodynamic changes needed to keep his cars tracking steadily at warp speeds. Professor Kamm's aerodynamic theories were a frequent reference – but really trial and error, using crudely attached body panels and plaster forms to speed the process, would eventually be the deciding factor. As mentioned earlier, the last GTO test car was run by Stirling Moss at Monza. Nicknamed, 'the Anteater,' it began to resemble the final GTO design with its extended nose designed to prevent high speed lift.

When *3223 GT*, the first GTO appeared, it still shared a lot of elements with its SWB predecessors. The shortened wheelbase, at 2400 mm, was unchanged – and the same basic construction continued, although the chassis tubing was rearranged somewhat.

The GTO was lighter and lower than the SWB and its engine was located back further in the reinforced chassis. The suspension, although similar in concept, was also modified. Under the long, low hood was a 250GT motor, hooked to a newly-designed five-speed gearbox. The engine, fitted with a dry sump to increase its oil capacity and to permit lower mounting and thus a lower centre of gravity, featured six Weber 38 DCN twin-choke carburettors, over-sized valves, needle bearing rocker arms and a lot of light-alloy castings. Just over 300 bhp was shown on factory data sheets.

The GTO's new body only partially resembled the SWB's. It was sleeker, longer, prettier and much more efficient at high speeds. As well, the factory designers

*In late 1964, the Lusso gave way to the 275GTB, a 3.3-litre berlinetta with racy lines reminiscent of the SWB's and GTO's. Special racing GTB/Cs had some competitive successes, such as the Maranello Concessionaries Le Mans GT win in 1966, but Ferrari's focus was increasingly on the prototype class with mid-engined cars like the 250LM. Shown is GTB 08801, a late long-nosed, 6-carburettor, torque-tube car belonging to the author. Photo Alan Weitz*

had improved on the SWB's, rugged, purposeful good looks, making the GTO even more authoritative. The GTO proceeded to win the GT Championships from 1962–64, until the rules changed once again.

With the GTO's limited numbers, Ferrari needed a street successor to the steel-bodied Lusso SWBs. A 'cooking' GTO would have run contrary to the car's design principles – and privateers, denied the real thing, would have modified them for racing anyway. Ferrari introduced the 250GT Berlinetta Lusso, a lovely Pininfarina-designed GT car not intended for racing, although one of two of them found their way on to road courses at the hands of enthusiastic owners.

Instead, the Lusso, as it was popularly called, was a comfortable, luxurious tourer with elegant, flowing lines. Thus the SWB's place in the Ferrari family tree branched into two directions. Understandably, with racing development reaching new heights, Ferrari recognized the day of the true dual-purpose car was effectively over.

Always in short supply, the demand for GTOs was such that some racing stables, as we saw in chapter 6, tried to build their own copies on modified SWB chassis. The SWB, last of the truly competitive 'drive it to the track, run it and drive home' berlinettas, was gone, but not forgotten. As time passed and GTO values shot up to unaffordable heights, people remembered the plucky SWBs. Today, a good steel Lusso is worth more than £300,000 – I've even seen some with competition features like the outside filler cap, etc, for more than £450,000.

Competition alloy cars, particularly the highly desirable SEFAC Comp/61s, today approach £500,000 in the late 1990s – no finer tribute to a great automobile.

# Specifications

As frequently mentioned in the text, SWB specifications often changed as the series developed. The appendix section on SWB chassis numbers indicates just a few of the variations.

### Engine
3-litre: 2953.22 cc, bore and stroke 73 × 58.8 mm. V12 cylinder, 60 degree, two single overhead camshafts, one for each bank of cylinders. Compression ratio varied from 9.2, 9.3:1 for Lusso cars to 9.3, 9.5:1 for 'normal' competition cars. Comp./61 engines had 9.7:1. Engine block was Silumin alloy. Late Testa Rossa cylinder heads and 7-bearing crankshaft. Twin Marelli distributors. Carburation varied with street cars utilizing three Weber twin-choke DCL 6 or three Weber DCL3s. *2111 GT* ordinarily had Solex C 40 PAA L carburettors. Competition berlinettas had three Weber 40 DCL 6 instruments. Comp./61 SWBs used Weber 46 DCL 3 or 46 DCF 3. Wet sump, 9-litre capacity (although some rebodied cars were fitted with dry sumps). Horsepower varied: street cars: 220–240 bhp at 7000 rpm, Competition: 260–275 bhp at 7000 rpm, Comp./61: 280 bhp plus at 7000 rpm. Approximate torque for 'normal' competition cars is estimated at 183 ft/lb at 5500 rpm; for Comp./61 cars, it's 203 ft/lb at 5500 rpm. Note: detail engine changes were often made during the production run from late 1959–1962.

### Transmission
Clutch: Fitchel & Sachs, single dry plate, cable-operated.

Gearbox: four-speed with Porsche patent synchromesh manufactured by Ferrari. Steel-bodied cars had a cast-iron case; Competition cars used a finned alloy case. An electric overdrive was tested in *2111 GT*, but it was not adopted.

Final drive: Rigid axle with ZF patented limited-slip differential manufactured by Ferrari. Final drive ratio with 7/32 at its lowest to 9/31 at its highest, or 4.85:1 and 3.44:1 respectively. Seven variations were available.

### Chassis
Multi-tube welded steel ladder type.

Front suspension: Double wishbones, coil springs, Miletto (early cars) or Koni tubular shock absorbers, 15 mm anti-roll bar, suspension height was adjustable.

Rear suspension: Semi-elliptic leaf springs, two torque arms each side, Miletto or Koni tubular hydraulic shock absorbers. Note: competition cars used stiffer, polished springs and harder shock settings than Lussos.

Steering: ZF worm and sector with 17:1 ratio. The last SWB, *4065 GT*, had a 20:1 ratio like the succeeding 250GT Berlinetta Lussos. Turning ratio was 48 ft, right and 37 ft, left. Number of turns was 1.5, lock to lock. Right-hand steering was not available until *1993 GT*. In all, only 14 cars were built with rh steering, and the prototype, *1539 GT*, was converted from lhd to rhd but is now lhd. Brakes: Dunlop four-wheel solid discs with cast-iron or light alloy calipers. 490 sq. in. of braking area. Lussos had Bendix power assist. Competition cars had a booster for the front brakes only.

### Wheels and tyres:

Borrani wire spoke wheels with 42 mm Rudge Whitworth hubs in various sizes, painted or chromed. Most cars used 400 × 5½ 16 in. wheels. Fifteen inch wheels were available in 1961. Sizes available included: 185 × 400 RW 3598, 400 × 5½ RW 3661, 185 × 15 RW 3711, 185 × 16 RW 3526/3264, 185 × 15 RW 3591, 6.50 × 16 RW 3687 and 6.00 × 15 RW 3690. Street cars were fitted with Pirelli Cinturatos; racers used Dunlop R-types. One special-bodied SWB, *1739 GT*, was equipped with Campagnolo solid alloys.

### Dimensions

Wheelbase: 2400 mm (94.4 in.)
Front track: 1354 mm (53.3 in.)
Rear track: 1349 mm (53.1 in.)
Net weight: Street cars approximately 2560–2649 lb, competition cars 2318–2406 lb.

### Bodywork – bodies were built up from many small pieces welded together.

Two-door berlinetta designed by Pininfarina and built by Scaglietti. Competition cars were all alloy, Lussos had steel bodywork. Frequently, Lusso decklids, doors and hoods were made of aluminium alloy. A two or sometimes three-piece belly pan ran from the radiator to the fuel tank edge. Integral cold-air type scoops mated to a carburettor cold–air box were built-in on most competition cars and some Lusso SWBs. Right-hand specifications are competition cars.
Overall length: 4153 mm (163.5 in.) (4200 mm/165.48 in.)
Overall width: 1651 mm (65.0 in.) (1720 mm/67.76 in.)
Overall height: 1283 mm (50.5 in.) (1270 mm/50.03 in.)
**Performance:** Theoretical top speed with 9/31 axle and 16 in. wheels was 166.6 mph, but it's unlikely an SWB could attain anywhere near this velocity due to aerodynamic limitations. With the 7/32 ratio and 16 in. wheels, top speed was 125.5 mph. Maximum practical speed for Lussos was 145–150, for competition cars, 150–155.
0–60 mph, standing start: 6.3 sec. (has been quoted).
Standing ¼ mile: 14.3 sec. at 105.0 mph.

# Production record

**Courtesy Stan Nowak, Jess Pourret and**
*Cavallino* (with modifications by the author)

*Below is a list of all known 250GT short wheelbase serial numbers, which also denotes whether the car was a race model or street car, and which provides a few comments on those cars that deserve a mention. This list is as accurate as possible at the moment, but use with caution,*

*since some SWBs haven't been found, some others have had their numbers changed, and some may not be SWBs at all. We wish to thank Alan Boe and Jess Pourret for their researches into the 250GT SWB chassis numbers.*

| | Serial No. | Type | Comments |
|---|---|---|---|
| **1959** | 1539 | Competition | The first SWB, 1959 Paris Salon car. |
| | 1613 | Competition | |
| | 1739 | Competition | Bertone body, special design. |
| **1960** | 1741 | Competition | The first 1960 SWB |
| | 1757 | Competition | Excellent GT racing record. |
| | 1759 | Competition | |
| | 1771 | Competition | |
| | 1773 | Competition | |
| | 1785 | Competition | |
| | 1791 | Competition | |
| | 1807 | Competition | 1960 Spa winner; good GT racing record. |
| | 1811 | Competition | Good hillclimb record. |
| | 1813 | Competition | |
| | 1849 | Competition | |
| | 1875 | Competition | Excellent GT racing record. New body. |
| | 1887 | Competition | |
| | 1905 | Competition | |
| | 1917 | Competition | 1962 Nurburgring 1000 GT winner. |
| | 1931 | Competition | 1961 Sebring GT winner. |
| | 1953 | Competition | |
| | 1965 | Competition | |
| | 1993 | Street | First steel SWB, and first rhd. |
| | 1995 | Street | rhd. |
| | 1997 | Competition | Good GT racing record. |
| | 1999 | Competition | 1960 Coppa Inter Europa winner. |
| | 2001 | Competition | 1960 Le Mans GT winner. |
| | 2009 | Competition | |
| | 2021 | Competition | |
| | 2025 | Competition | |
| | 2033 | Competition | |
| | 2035 | Competition | Excellent rallying record. |
| | 2053 | Competition | Used for experiments for upcoming GTO; Drogo body. |
| | 2055 | Street | |
| | 2067 | Street | Drogo body added on later. |
| | 2069 | Street | |
| | 2083 | Competition | |
| | 2095 | Competition | |
| | 2111 | Street | Test car for 1961 SWBs; special engine. |
| | 2119 | Competition | First Walker/Moss SWB; many firsts, including 1960 Tourist Trophy. RHD. |
| | 2127 | Competition | |
| | 2129 | Competition | 1960 Tour de France winner. |
| | 2141 | Competition | |
| | 2149 | Competition | 1960 Montlhéry 1000 winner; later converted to street version. |
| | 2159 | Competition | Good rally and hillclimb record. |
| | 2163 | Competition | 1961 Nurburgring 1000 GT winner. |
| | 2165 | Competition | |
| | 2177 | Competition | With steel body, used for hillclimbs. |
| | 2179 | Competition | |
| | 2209 | Competition | Drogo body added on later. |
| | 2221 | Street | rhd. Some comp. mods |
| | 2231 | Competition | Modified twin front grill. |
| | 2237 | Competition | |

| | | |
|---|---|---|
| 2243 | Street | |
| 2251 | Street | |
| 2265 | Street | |
| 2269 | Street | |
| 2283 | Street | Vents on sail panels. |
| 2289 | Street | |
| 2291 | Street | |
| 2321 | Competition | |
| 2335 | Street | rhd. |
| 2347 | Street | |
| 2389 | Street | The last 1960 SWB. |
| **1961** 2399 | Street | The first 1961 SWB. |
| 2417 | Competition | 1961 Spa winner; Comp./61 engine; good race record. |
| 2419 | Street | |
| 2429 | Street | Pininfarina 400 SA body; experimental; sold to special customer; Comp./61 engine. |
| 2439 | Competition | Comp./61 engine. |
| 2443 | Street | |
| 2445 | Competition | Comp./61 engine; excellent hillclimb record. |
| 2455 | Competition | |
| 2491 | Street | Zagato body. |
| 2501 | Street | |
| 2521 | Street | |
| 2549 | Street | |
| 2551 | Street | |
| 2563 | Street | |
| 2589 | Street | |
| 2595 | Street | |
| 2613 | Street | Pininfarina 400SA body |
| 2617 | Street | |
| 2639 | Street | Faired-in headlights added later. |
| 2643 | Competition | Pininfarina 400SA body; GTO 'prototype;' Comp./61 engine. |

| | | |
|---|---|---|
| 2649 | Street | Covered headlights added later. |
| 2667 | Competition | |
| 2669 | Street | |
| 2687 | Competition | Comp./61 engine; excellent GT racing record. |
| 2689 | Competition | 1961 Le Mans GT winner; Comp./61 engine. |
| 2701 | Competition | Factory test car; Comp./61 engine. |
| 2725 | Competition | Comp./61 engine. |
| 2729 | Competition | Comp./61 engine. |
| 2731 | Competition | Comp./61 engine. |
| 2733 | Competition | Comp./61 engine; decent GT race record. |
| 2735 | Competition | Second Walker/ Moss SWB; excellent race record; Comp./61 engine; later Drogo body added; rhd. |
| 2765 | Street | |
| 2767 | Competition | Comp./61 engine. |
| 2787 | Competition | Comp./61 engine; excellent GT race record. |
| 2807 | Competition | Decent rally and hillclimb record. |
| 2819 | Competition | Decent race record; Bizzarini had Neri e Bonaccini breadvan body added later; Comp./61 engine. |
| 2839 | Competition | Later body added. |
| 2845 | Competition | Comp./61 engine. |
| 2863 | Street | |
| 2909 | Street | |
| 2917 | Street | |
| 2935 | Street | |
| 2937 | Competition | 1961 Tour de France winner; Comp./61 engine. |

119

| | | | |
|---|---|---|---|
| 2939 | Competition | 1962 Spa winner; Comp./61 engine; good rally record. | |
| 2973 | Competition | 1962 Tour de France winner; Comp./61 engine. | |
| 2985 | Street | | |
| 2989 | Street | | |
| 3005 | Competition | 1961 Montlhéry/1000 winner; Comp./61 engine. | |
| 3035 | Street | | |
| 3037 | Street | rhd. | |
| 3039 | Street | | |
| 3067 | Street | rhd. | |
| 3073 | Street | | |
| 3087 | Street | | |
| 3107 | Street | | |
| 3113 | Street | | |
| 3129 | Street | | |
| 3143 | Competition | With steel body used for hillclimbs. | |
| 3169 | Street | | |
| 3175 | Street | Used for GTO studies | |
| 3233 | Street | | |
| 3269 | Street | Bertone body, special. | |
| 3281 | Street | rhd. | |
| 3287 | Street | rhd. | |
| 3315 | Street | | |
| 3327 | Competition | | |
| 3331 | Street | Last 1961 SWB | |
| **1962** 3337 | Street | First 1962 SWB | |
| 3359 | Street | Rear modified later. | |
| 3367 | Street | | |
| 3379 | Street | | |
| 3401 | Street | | |

| | | |
|---|---|---|
| 3409 | Street | |
| 3425 | Street | |
| 3431 | Street | |
| 3463 | Street | First SWB with 1962 design features; rhd. |
| 3477 | Street | |
| 3487 | Street | |
| 3507 | Street | |
| 3539 | Competition | |
| 3551 | Street | rhd. |
| 3565 | Competition | |
| 3577 | Street | |
| 3605 | Street | rhd. |
| 3615 | Street | Pininfarina 400SA body |
| 3639 | Street | |
| 3657 | Street | rhd. |
| 3695 | Street | |
| 3709 | Street | |
| 3735 | Street | |
| 3771 | Street | Neri e Bonaccini body (later), Nembo spyder. |
| 3815 | Street | |
| 3829 | Street | |
| 3847 | Street | |
| 3863 | Street | |
| 3877 | Street | |
| 3911 | Street | |
| 3963 | Street | |
| 4037 | Street | |
| 4051 | Street | |
| 4057 | Street | |
| 4065 | Street | Last of the SWBs made. |

Both 1999 and 2687 exist today but are not 250GT SWBs in anything but name.

# Competition

Courtesy Alan Boe and *The Prancing Horse*, and Jess Pourret

| Event | Date | Driver(s) | Chassis number | Race results |
|---|---|---|---|---|
| 12 Hours of Sebring, Fla. | 3/26/60 | Hugus/Pabst | 1785 | 4th overall |
| | | Reed/Connell | 1741 | 5th overall |
| | | Sturgis/D'Orey | 1539 | 6th overall |
| | | Arents/Kimberly | 1773 | |
| 44th Targa Florio | 5/8/60 | Lenza/Maglione | 1791 | 9th overall |
| | | Le Pira/Serini | 1875 | 12th overall |
| | | Ferraro/Zampero | 1813 | 18th overall |
| Spa, Belgium | 5/60 | Mairesse | 1807 | 1st overall |
| 1000 km of the Nurburgring | 5/22/60 | Beurlys/Noblet | 1811 | 15th overall |
| 8th Grand Prix of Rouen, France | 6/12/60 | Seidel | 1807 | 3rd overall |
| | | Dumay | 2127 | 4th overall |
| 24 Hours of LeMans, France | 6/25–26/60 | Tavano/Dumay | 2001 | 4th overall, 1st GT, 105 mph |
| | | Arents/Connell | 1931 | 5th overall, 2nd GT, 104 mph |
| | | 'Elde'/Noblet | 2021 | 6th overall, 3rd GT, 104 mph |
| | | Hugus/Pabst | 1759 | 7th overall, 4th GT, 104 mph |
| 25th Tourist Trophy, Goodwood, England | 8/20/60 | Moss | 2119 | 1st overall |
| | | Whitehead | 2009 | 5th overall |
| Redex Trophy Race, Brands Hatch, England | 8/27/60 | Moss | 2119 | 1st overall |
| Coppa Inter Europa, Monza, Italy | 9/60 | Abate | 1999 | 1st overall |
| | | Guichet | 1887 | 2nd overall |
| | | Tuselli | 2053 | 3rd overall |
| Tour de France | 9/60 | Mairesse/Berger | 2129 | 1st overall |
| | | Schlesser/ Dumay | 2127 | 2nd overall |
| | | Tavano/Martin | 2001 | 3rd overall |
| 1000 km of Montlhéry, Paris, France | 10/23/60 | Bianchi/Gendebien | 2149 | 1st overall |
| | | Mairesse/Von Trips | | 2nd overall |
| | | Schlesser/Simon | 2209 | 2nd overall |
| | | Dumay/Tavano | 2127 | 3rd overall |
| | | Whitehead/Taylor | 2009 | 5th overall |
| Nassau, Bahamas | 12/60 | Moss | 2119 | 1st overall |
| 12 Hours of Sebring, Fla. | 3/25/61 | McCluggage/Eager | 1931 | 10th overall, 1st GT |

| | | | |
|---|---|---|---|
| Snetterton, England | 3/25/61 | Parkes | 2119 | 1st overall |
| | | Whitehead | 2009 | 2nd overall |
| Fordwater Trophy Race, Goodwood, England | 4/3/61 | Parkes | 2119 | 1st overall |
| | | Whitehead | 2009 | 4th overall |
| Oulton Park, England | 4/15/61 | Sears | 2119 | 4th overall |
| | | Whitehead | 2009 | 5th overall |
| Spa, Belgium | 5/14/61 | Mairesse | 2417 | 1st overall |
| | | Whitehead | 2009 | 3rd overall |
| | | Noblet | 2021 | 4th overall |
| | | Andersson | 2439 | 5th overall |
| 1000 km of the Nurburgring | 5/28/61 | Abate/Davis | 2163 | 4th overall, 1st GT |
| | | Mairesse/Baghetti | 2417 | 5th overall, 2nd GT |
| | | Felder/Nocker | 1917 | 12th overall |
| | | Berger/Beurlys | 1965 | 16th overall |
| 24 Hours of LeMans, France | 6/10– 11/61 | Noblet/Guichet | 2689 | 3rd overall, 1st GT, 110 mph |
| | | Grossman/Pilette | 2731 | 6th overall, 2nd GT, 107 mph |
| British Empire Trophy, GT cars, Silverstone, England | 7/8/61 | Moss | 2735 | 1st overall, 1st GT |
| | | Whitehead | 2009 | 4th overall, 4th GT |
| Six Hours of Auvergne, France | 7/9/61 | Mairesse | ? | 1st overall |
| GT Race, Nurburgring, Germany | 8/6/61 | Abate | 2733 | 1st overall |
| Peco Trophy Race, Brands Hatch, England | 8/7/61 | Moss | 2735 | 1st overall |
| Pescara Grand Prix, Pescara, Italy | 8/15/61 | Arents/Hamill | 2725 | 4th overall, 1st GT |
| | | Bettoia/'Kim' | 2767 | 6th overall, 2nd GT |
| | | Cacciari/Bertocco | 2095 | 7th overall, 3rd GT |
| Tourist Trophy, Goodwood, England | 8/19/61 | Moss | 2735 | 1st overall, 86.62 mph |
| | | Parkes | 2119 | 2nd overall |
| Coppa Inter-Europa, Monza, Italy | 9/10/61 | Noblet | 2689 | 1st overall, 110 mph |
| | | Lualdi | 2687 | 3rd overall |
| Tour de France | 9/61 | Mairesse/Berger | 2937 | 1st overall |
| | | Gendebien/Bianchi | 2819 | 2nd overall |
| | | Trintignant/Cavrois | 2845 | 3rd overall |
| | | Berney/Gretener | 2939 | 4th overall |
| Molyslip Trophy Race, Snetterton, England | 10/3/61 | Parkes | 2119 | 1st overall, 92.4 mph |

| | | | | |
|---|---|---|---|---|
| 1000 km of Montlhéry, Paris, France | 10/22/61 | Rodriguez/Rodriguez | 3005 | 1st overall, 1st GT |
| | | Mairesse/Bianchi | 2937 | 2nd overall, 2nd GT |
| | | Vacarella/Trintignant | 2819 | 3rd overall |
| | | Dumay/Schlesser | 2729 | 4th overall |
| | | Abate/Davis | 2767 | 5th overall |
| Nassau Tourist Race, GT cars, Nassau, Bahamas | 12/4/61 | Moss | 2735 | 1st overall, 80.113 mph |
| | | Grossman | ? | 2nd overall |
| | | Hathaway | 1773 | 3rd overall |
| | | Hayes | 2237 | 4th overall |
| Daytona Beach, Fla., SCCA National Race | 1/28/62 | Thiem | 3005 | 6th overall, 1st AP |
| 12 Hours of Sebring, Fla. | 3/24/62 | Hamill/Serena | 2725 | 4th overall, 2nd GT |
| | | Hugus/Reed | ? | 8th overall, 4th GT |
| | | Dietrich/Haas/ Hayes | 3327 | 32nd overall |
| Marlboro Governor's Cup, SCCA National Race, Marlboro Md. | 4/15/62 | Grossman | 2731 | 1st AP, 64.08 mph |
| | | Thiem | 3005 | 3rd AP |
| 46th Targa Florio, Sicily | 5/6/62 | Rolland/ DeLageneste | 2807 | 5th overall, 2nd GT |
| | | Crispi | 1813 | 16th overall |
| SCCA National Race, Cumberland, Md. | 5/13/62 | Grossman | 2731 | 2nd overall, 2nd AP |
| | | Hayes | 2237 | 3rd overall, 3rd AP |
| Spa, Belgium | 5/62 | Berney | 2939 | 1st overall |
| | | Noblet | 2689 | 2nd overall |
| | | Berger | 2937 | 3rd overall |
| | | Crevits | 2445 | 4th overall |
| 1000 km of the Nurburgring, Germany | 5/27/62 | Nocker/Seidel | 1917 | 5th overall, 1st GT |
| | | Noblet/Guichet | 2689 | 7th overall |
| | | Oreiller/De la Geneste | 2787 | 10th overall |
| SCAA National Race, Stuttgart, Ark. | 5/27/62 | Thiem | 3005 | 2nd overall, 2nd AP |
| June Sprints, SCCA National Race, Elkhart Lake, Wis. | 6/16/62 | Dietrich | 3327 | 2nd overall, 2nd AP |
| | | Thiem | 3005 | 3rd overall, 3rd AP |
| Six Hours of Marlboro, Marlboro, Md. | 6/17/62 | Hayes/G. Hobbs | 2237 | 1st overall, 56.4 mph |
| | | Wyllie/Georgi | 1931 | 2nd overall |
| SCCA National Race, Lime Rock, Conn. | 6/30/62 | Georgi | 1931 | 1st overall, 1st AP |
| | | Grossman | 2731 | 2nd overall, 2nd AP |
| SCAA National Race, Lake Garnett, Kan. | 7/8/62 | Thiem | 3005 | 2nd overall, 2nd AP |

| | | | | |
|---|---|---|---|---|
| SCCA National Race, Thompson, Conn. | 9/3/62 | Georgi | 1931 | 1st overall, 1st AP |
| | | Thiem | 3005 | 2nd overall, 2nd AP |
| | | Grossman | 2731 | 3rd overall, 3rd AP |
| SCCA National Race, Elkhart Lake, Wis. | 9/8/62 | Thiem | 3005 | 1st overall, 1st AP |
| Tour de France | 9/62 | Simon/Dupeyron | 2973 | 1st overall |
| | | De la Geneste/Burglin | 2807 | 5th overall |
| 1000 km of Montlhéry, Paris, France | 10/21/62 | Davis/Scarfiotti | 2819 | 3rd overall |
| | | Simon/Berger | 2973 | 6th overall |
| Nassau Trophy, Nassau, Bahamas | 12/9/62 | Hayes | 2237 | 5th overall |

# Special-bodied SWBs

## Original special-bodied SWBs

*1739*  The first Bertone design
*2429*  Superamerica-style by Pininfarina – enclosed lights
*2613*  Pininfarina SA-style coupé – open lights
*2643*  250GTO prototype by Pininfarina
*3269*  The second Bertone design
*3615*  Pininfarina 400 SA-style coupé – open lights

## Rebodied SWBs

*2053*  Drogo Berlinetta (destroyed)
*2067*  Drogo rebody similar to 2209
*2209*  Tadini-designed Drogo rebody
*2491*  Zagato spyder
*2735*  Drogo Berlinetta (originally Walker-Moss Comp./61)
*2819*  Neri e Bonaccini 'breadvan'
*3771*  Neri e Bonaccini Spyder body – later Nembo

## Modified SWBs

*2111*  1960 SWB re-bodied as '61–'62 prototype
*2231*  Twin-nostril 'Carlo Chiti' nose
*2639*  Faired-in headlights and GTB taillights added later
2649  Covered headlights, flush taillights, other minor modifications
*2839*  Custom body modifications, resembles a 275GTB
*3359*  Lusso-275GTB taillights

*(All with GT suffix)*

# Acknowledgements

I'd like to thank the many people who helped me so much with the preparation of this book. My sincere thanks go, first and foremost, to Alan Boe, a true authority on SWBs, for his generous help with the manuscript and with many photographs. I'd also like to acknowledge the fine work of Jess G. Pourret whose book, *The 250 GT Competition*, remains the standard for any work on SWBs. I'm grateful to Dean Batchelor, Karl Dedolph and Jonathan Thompson who read the manuscript and helped with many photographs.

The following people also contributed information and/or photographs: Marc Tauber, Andre Dudek, Stan Nowak, Peter Sachs, Anatoly Arutunoff, Gerald Rousch, Alan Weitz, Denise McCluggage, David Seielstad, Tim Parker, Pierre Bardinon, Marcel Massini, Dr. Mel Wilner, Lorenzo Zambrano, Marshall Matthews, Jacques Vaucher, Mike Gourley, Gary Schonwald, Chris Leyden, Lowell Paddock, John Risch, Steve Tillack, Robert Ames, George Murtha, Charles Betz, Fred Peters, Bob Dunsmore, Jerry Gamez, Charles Reid, Dr. Phillip Bronner, Kurt Miska, Jerry Lynch, Robert Stirnkorb, George Heiser, Autopresse International, and Chris and Jeremy Gross.

I speak for all of us in thanking Enzo Ferrari, for creating these wonderful cars we all love. Finally, cheers to my old friend Ra, who patiently typed the manuscript. To all of you, many thanks.

# INDEX

Abarth   41, 79
Abate, Carlo   81, 91, 94
AC Cobra   113
Aga Khan   13
Alabama   7
*Altissimo*   35, 36, 58, 59, 60
Andersson, Gunnar   86
Angola GP   26, 86
Arents, George   91, 93
Aruntoff, Anatoly   77, 78
Aston Martin   98, 113
*Autocar*   104, 106
*Automobile Quarterly*   6
*Autosport*   99, 100

Bardinon, Pierre   97
Beasley, Ron   109
Bendix   40, 41
Berger, Tojo   91, 97, 111
Bertone, Nuccio   56, 74, 75, 79
Beurlys   88
Bianchi, Lucien   91, 111
Bizzarini, Giotto   26, 81, 84, 85
Boe, Alan   36, 49, 54, 60
Bolster, John   99, 100
Bonnier, Jo   37
Borrani   23, 43, 75, 78, 79, 84
Brands Hatch   81, 89, 90
Bridgehampton, LI   95
Brunn, Sigi   61
Bugatti, Ettore   37
Burton, George   105

Campagnolo   24, 74, 75
*Car & Car Conversions*   105
Cavrois   37
Chevrolet Corvette   95
Chinetti, Luigi   77, 78, 88, 93, 98
Chippendale   105
Chiti, Carlo   26, 85, 87
Conan Doyle, Adrian   87
Connell, Alan   91, 93
Constructor's Championship   54
Coppa Inter-Europe   90
Cumberland, MD   96

Daytona Continental   90
Dedolph, Karl   19, 21, 22, 23, 24
Dijon   106
Drojo, Piero   83, 84, 85
Dron, Tony   109
Dumay   88, 91
Dunlop   24, 25, 40, 108

Eager, Allen   91, 92, 93, 97
Elektron   55
Englebert   24
Everflex   22

Fehlman, Herr   100
Ferrari
   250GT Berlinetta Lusso   27
   250GT Competition   20
   250 Europa GT   10
   250LM   21
   250GTO   9, 21, 49, 82, 89, 111, 112, 113
   250TDF Berlinetta   9, 10, 19, 22
   275GTB   21, 49, 82, 114
   1321 GT   24
   1357 GT   15
   1377 GT   20
   1451 GT   16
   1461 GT   20
   1505 GT   10
   1521 GT   19, 21
   1523 GT   49
   1959 GT   15
   'Anteater'   114
   Cabriolet Speciale   83
   California Spyder   10, 13, 16, 49, 50, 83
   400SA Spyder   81
   308 Quattrovalvole   110
   Series I Pinin Farina Spyders   13
   Series II Pinin Farina 250 Cabriolet   13
   Superamerica   76, 77, 78, 84, 87
   Testa Rossa   8, 15, 46, 55, 83, 98, 99, 111

Engines:
   128 F   46
   128 DF   45, 46
   168   45
   168 B   45
Ferrari, Enzo   8, 16, 20, 45, 55, 79, 111, 113
*Ferrari Market Letter*   8
FIA   40, 111, 112, 113
Fispa   46, 53
Fitzgerald, Warren   20, 84
Forghieri, Mauro   26

Gardini   85
Gatta, F.   79
Gendebien, Olivier   22, 85, 91
Geneva Salon   15, 25, 26, 79, 83
Genuizzi, Edoardo   105
Germany   92
Goodwood   89, 90, 99, 109
Grayson, Stan   89
Grey   88
Grossman, Bob   16, 93, 95, 97
Guichet, Jean   70

Harrah, Bill   13
Hill, Graham   37
Hill, Phil   110
Hoare, Col. Ronnie   99
Houdaille   23, 40
Hugus, Ed   28, 88, 100

Imola   105
Ireland   86
Italy   74, 92

Jaeger, Professor   114
Jaguar   40, 113
Johnson, Jr., Charles   84

Kamm, Professor   114
Karussel   92
Kerrison, Chris   83, 86
Koni   39, 40, 44
Kreisler, Charlie   100

# INDEX

Laguna Seca  **21, 96**
Lake, Bill  **87**
Le Mans  **16, 18, 20, 23, 24, 28, 39, 41, 70, 77, 85, 88, 89, 91, 92, 93, 107**
Lenza  **88**
Leyden, Chris  **38**
Lime Rock Park  **103**
London Show, 1961  **78**
Lucas  **105**
Lucchini, Giuseppe  **102**

Maglione  **88**
Mairesse, Willy  **90, 97**
Maranello  **26, 45, 55, 68, 73, 80, 112**
  Concessionaires  **99, 114**
Marchal  **49, 62**
Marelli  **46, 53**
Marlboro  **93, 94**
Martin  **91**
Maserati  **98**
McCluggage, Denise  **91, 92, 93**
Meade, Tom  **82, 84**
Meadowdale, Il  **95**
Merrit, Dick  **20, 84**
Michelotti, Giovanni  **74**
Miletto  **39, 40, 44**
Mini  **97**
Modena  **17, 39, 97**
Montery Historic Races  **96**
Montlhéry 1000 Km  **37, 91, 109**
Monza  **17, 22, 35, 41, 58, 90, 114**
Moss, Stirling  **80, 86, 89, 114**

Nassau  **27, 89, 91**
  Trophy  **91**
Neri e Bonaccini  **84, 85, 86**
Noblet, Pierre  **70, 88, 98**
Nocker  **104**
Nowak, Stan  **36**
Nurburgring  **88, 92, 97, 104, 106**

Pabst, Augie  **28, 88, 100**
Paddock, Lowell  **6**
Paris Salon  **10**
  1954  **13**
  1959  **25**
Parkes, Mike  **98**
Pearse, Roy  **109**
Pebble Beach  **80**
Philips, Simon  **107**
Pininfarina  **16, 21, 23, 26, 35, 54, 65, 74, 76, 77, 78, 81, 83, 99, 100**
Porsche  **40, 109, 113**
Pourret, Jess G.  **20, 24, 36, 37, 43, 49, 60, 76, 77, 84, 85, 86**

Redex Trophy  **89**
Risch, John  **46, 59**
*Road & Track*  **110**
Road America  **27**
Rouen  **88**
Rousch, Gerald  **7, 8**
Rubirosa, Porfirio  **13**
Rudge  **43**

Sachs, Peter  **81, 83, 84**
Sarthe  **107**
Scaglietti, Sergio  **13, 17, 20, 23, 25, 27, 31, 59, 65, 87, 97**
Schaevitz, Gary  **38**
Schlesser, Jo  **39, 91**
Schonwald, Gary  **38**
SCI  **101**
Scuderia Serenissima  **37, 82, 94, 109**
Scuderia National Belge  **86**
Sebring  **27, 77, 88, 92, 97, 98**
Seidel  **88, 90, 104**
SEFAC  **47, 53, 54, 65, 77, 91, 115**
Silverstone  **55, 96, 108**
Simon  **91**
Snap  **98**
Snetterton  **48**
Solex  **48**
Spa  **82, 86**

*Sports Car Illustrated*  **100, 102, 105**
Sturgis  **88**

Tardini  **84**
Targa Florio  **88, 89**
Taruffi  **97**
Tatum, Gordon  **85**
Tauber, Marc  **8**
Tavano, Ferdinand  **16, 88, 91**
Tavoni  **85**
Thiem, D  **94**
Thompson (CT)  **93, 97**
Thompson, Dick  **93**
*Thoroughbred & Classic Car*  **107, 109**
Tillack, Steve  **74, 76, 79, 80**
Tour de France, 1959  **24, 85**
Tourist Trophy  **89**
Transport, Minister of  **106**
Trintignant, Maurice  **37**
Turin Show  **25, 26**
  1974  **77, 78**

USA  **74**

Vaccari  **39**
Vallaster, Ado  **29**
Vaucher, Jacques  **25**
Volpi, Count  **13, 82, 85**

Walker, Rob  **80, 86, 89, 90, 91**
Weber  **45, 47, 48, 49, 55, 77, 85, 86, 97, 102, 115**
Whitehead, Graham  **107**
Whitehead, Peter  **98**

Yenko, Don  **95**

Zagato  **13, 78, 98, 113**
Zambrano  **79**
ZF  **40**